Radiation Protection in Medical Physics

NATO Science for Peace and Security Series

This Series presents the results of scientific meetings supported under the NATO Programme: Science for Peace and Security (SPS).

The NATO SPS Programme supports meetings in the following Key Priority areas: (1) Defence Against Terrorism; (2) Countering other Threats to Security and (3) NATO, Partner and Mediterranean Dialogue Country Priorities. The types of meeting supported are generally "Advanced Study Institutes" and "Advanced Research Workshops". The NATO SPS Series collects together the results of these meetings. The meetings are coorganized by scientists from NATO countries and scientists from NATO's "Partner" or "Mediterranean Dialogue" countries. The observations and recommendations made at the meetings, as well as the contents of the volumes in the Series, reflect those of participants and contributors only; they should not necessarily be regarded as reflecting NATO views or policy.

Advanced Study Institutes (ASI) are high-level tutorial courses intended to convey the latest developments in a subject to an advanced-level audience

Advanced Research Workshops (ARW) are expert meetings where an intense but informal exchange of views at the frontiers of a subject aims at identifying directions for future action

Following a transformation of the programme in 2006 the Series has been re-named and re-organised. Recent volumes on topics not related to security, which result from meetings supported under the programme earlier, may be found in the NATO Science Series.

The Series is published by IOS Press, Amsterdam, and Springer, Dordrecht, in conjunction with the NATO Public Diplomacy Division.

Sub-Series

A.	Chemistry and Biology	Springer
B.	Physics and Biophysics	Springer
C.	Environmental Security	Springer
D.	Information and Communication Security	IOS Press
E.	Human and Societal Dynamics	IOS Press

http://www.nato.int/science
http://www.springer.com
http://www.iospress.nl

Series B: Physics and Biophysics

Radiation Protection in Medical Physics

Edited by

Yves Lemoigne

European Scientific Institute
Site d'Archamps
France

Alessandra Caner

European Scientific Institute
Site d'Archamps
France

Springer

Published in Cooperation with NATO Public Diplomacy Division

Proceedings of the NATO Advanced Study Institute on Radiation Protection
in Medical Physics Activities
Archamps, France
26 November – 1 December 2009

ISBN 978-94-007-0259-2 (PB)
ISBN 978-94-007-0246-2 (HB)
ISBN 978-94-007-0247-9 (e-book)

Published by Springer,
P.O. Box 17, 3300 AA Dordrecht, The Netherlands.

www.springer.com

Printed on acid-free paper

Preface

This book contains the lectures given at the NATO Advanced Study Institute ASI-983455 "Radiation Protection in Medical Physics Activities", held at the European Scientific Institute of Archamps (ESI, Archamps – France) from November 19 to November 24, 2009. The ASI course was structured in three parts: the first was dedicated to general radiation protection principles while the second shortly reviewed the radiobiology principles indispensable to operators and specialists in this field. The third part was dedicated to radioprotection implementation for medical physics activities in hospitals. The ASI took place after a 5 week period dedicated to the European School of Medical Physics (ESMP), which was devoted to medical imaging and radiotherapy. Being aware of the importance of radiation protection in hospital and medical physics activities, a number of ESMP participants chose to extend their stay and attend the ASI lectures too. The ASI courses devoted to nuclear medicine and digital imaging techniques have been collected in two volumes of the NATO Science Series entitled "Physics for Medical Imaging Applications" (ISBN 978-1-4020-5650-5) and "Molecular imaging: computer reconstruction and practice" (ISBN 978-1-4020-8751-6). The Radiotherapy and Brachytherapy ASI courses are available in a volume of the NATO Science Series entitled "Physics of Modern Radiotherapy & Brachytherapy" (ISBN 978-90-481-3096-2).

Every year in autumn ESI organises the European School of Medical Physics, which covers a large spectrum of topics ranging from Medical Imaging to Radiotherapy over a period of five weeks. Thanks to the Cooperative Science and Technology sub-programme of the NATO Science Division, a sixth week was added in 2009, structured as ASI courses dedicated to "Radiation Protection in Medical Physics Activities". This allowed the participation of experts and students from 20 different countries, with diverse cultural background and professional experience, all of whom could fruitfully share their professional experience and discuss open problems and issues.

This opportunity is particularly enriching for our colleagues from the Southern Mediterranean Basin (Algeria, Egypt, and Morocco) who can seldom profit of similar exchanges with the European scientific community.

A very pleasant surprise was the exceptional increase in the rate of participation of North African colleagues (30% of total) , which can most likely be ascribed to the active role played by the ASI co-director, Dean Jamal DERKAOUI from the Mohammed Premier University in Oujda, Morocco. A further positive outcome of NATO ASI support is the publication of this book, containing the lecture series contributed by speakers during the ASI.

We hope this book will become a reference in radioprotection, addressing an audience of young medical physicists everywhere in the world, increasingly sensitive to radioprotection in their medical physics activities at hospitals and radiotherapy facilities

We wish to thank all the participants, who allowed the ASI at Archamps to be a success within an excellent international atmosphere: lecturers, students (who participated actively) and all the ESI team (Manfred Buhler-Broglin, Alessandra Caner, Tamara Barberan, Filiz Demolis and Davide Vitè).

Many thanks to the Hôpital Cantonal de Genève and to the Radiation Control section of CERN, the European Centre for Nuclear Physics in Geneva, which allowed us to visit the radiation facilities they are in charge of and introduced us to the health safety measures implemented to protect several thousand staff members.

Finally, we wish to thank and express our gratitude to the Cooperative Science and Technology sub-programme of the NATO Science Division, lead by Professor Fausto Pedrazzini, without whom this Advance Study Institute would have not been possible.

Yves Lemoigne,[1] Co-Director of ASI-983455

[1]European School of Medical Physics, European Scientific Institute, Bâtiment Le Salève, Site d'Archamps F-74166 Archamps (France).

Contents

LIST OF ASI-983455 Contributors

Name		address
Dr. R. Calendrino	Lecturer	IRCCS San Raffaele, via Olgettina 60 - 20132 Milano, Italy
Dr. A. Caner	Proceedings Editor	ESI, Batiment Le Saleve, Site d'Archamps - 74160 Archamps, France
Dr. S. Christofides	Lecturer	Nicosia G. Hospital, Strovolos - 2029 Nicosia, Cyprus
Prof. Jamal DERKAOUI	ASI Co-Director	Université Mohamed 1er - 60000 Oujda, Morocco
Dr. P. Kukolowicz	Lecturer	Med. PhyS; Dept. Holycross Cancer Centre - 25-734 Kielce, Poland
Dr. Yves LEMOIGNE	ASI Co-Director	ESI, Batiment Le Saleve, Site d'Archamps - 74160 Archamps, France
Dr. C. Lewis	Lecturer	King's College Hospital, Denmark Hill - SE5 9RS London, UK
Prof. Jim Malone	Lecturer	Trinity College, St James Hospital - Dublin 8, Ireland
Dr. P. Peschke	Lecturer	DKFZ, Im Neuenheim Feld 280 - 69120 Heidelberg, Germany
Dr. R. Padovani	Lecturer	SO di Fisica Sanitaria, Ospedale S. Maria della Misericordia - 33100 Udine, Italy
Dr. M. Rehani	Lecturer	IAEA, P.O. Box 100, Wagramer Strasse 5 - A-1400 Vienna, Austria
Dr. N. Samet	Lecturer	SoS Hancesti 20, App 130 - 2028 Chisinau, Republic of Moldova
Dr. M.A. Staniszewska	Lecturer	Med. Imaging Dept. Medical University, Al Kosciuski - 90-419 Lodz, Poland

LIST OF ASI-983455 Participants: Directors, Students and Lecturers

Name	citizenship	Country	code	City	address
M. Hazem Abd El-Hafiz	Eg	**Egypt**		Cairo	Cairo University, 15, El-Galaa st., El-haram, Flat:7 Giiza, Cairo
Prof. Hoda Abou-Shady	Eg	**Egypt**		Cairo	Cairo University
M. Tamer Abuelfadl	Eg	**Egypt**		Cairo	Electronics & Communications Dept., Faculty of Engineering, Giza, Cairo
M. Etam Akbas	Tr	**Turkey**	0324	Mersin	Mersin University, Department of Medical Biology and Genetic
M. Loay Al Alfy	Eg	**Egypt**		Cairo	55, st#47 tahrir city, Imbaba
Ms Inna Apostolova	Bg	**Bulgaria**	1592	Sofia	zhk.Druzhba"1 bl.67 fl. 14 app.61
M. Ashraf Arafat	Eg	**Egypt**		Maddi Cairo	Atomic Energy Authority, 3, Abd El salam St, Tura El Esment, Maddi
Ms. T. Barberan	Fr	**France**	74160	Archamps	ESI, Batiment Le Saleve, Site d'Archamps
M. Luca Bellesi	It	**Italy**	00157	Roma	Università Cattolica del Sacro Cuore, Via Pio IX 154

(continued)

(continued)

Name	citizenship	Country	code	City	address
M. Ran Ben-Hur	IL	**Israel**	18/15	Petach Tikva	Rabin Medical Center, Ichilov
M. Azzeddine Benlamoudi	Dz	**Algeria**	5000	Batna	Cité 126 logts Bloc 5 N°42
M. Messaoud Boubir	Dz	**Algeria**	5000	Batna	university Hadj Lakhda
M. Abdelali Boughaba	Dz	**Algeria**	5000	Batna	Univesity of Batna, Avenue Mohamed El Hadi Boukhlouf
M. Abdesselem Brihmat	Dz	**Algeria**	25230	Hamma Bouziane	CHU Benbadis, Constantine, 134, Rue De L'ALN
Dr. R. Calendrino*	It	**Italy**	20132	Milano	IRCCS San Raffaele, via Olgettina 60
Dr. A. Caner	It	**France**	74160	Archamps	ESI, Batiment Le Saleve, Site d'Archamps
Dr. F. Carrasco	Pt	**Portugal**	4200-072	Porto	IPOPFG, Rua Dr António Bernardino de Almeida
Dr. S. Christofides*	UK	**Cyprus**	2029	Nicosia	Nicosia G. Hospital, Strovolos
Ms Tulin Çora	Tr	**Turkey**	42080	Konya	Selcuk Üniversity, Medical Faculty, Medical Genetics
M. Luis Cunha	Pt	**Portugal**	4200-072	Porto	IPOPFG, Rua Dr António Bernardino de Almeida
Ms Filiz Demolis	Tr	**France**	74160	Archamps	ESI, Batiment Le Saleve, Site d'Archamps
Prof. Jamal DERKAOUI,	Ma	**Morocco**	60 000	Oujda	Université Mohamed 1er
M. Ionut Dumitrache	Ro	**Romania**	032499	Bucharest	Str. J. Steriadi nr.14, bl.i17, ap.54, sect 3
Ms Gabriela Florescu	Ro	**Romania**	77125	Magurele-Ilfov	IFIN HH, 16, Fizicienilor street, M3, ap.7
Dr. Forkel-Wirth	Ch	**Switzerland**	1211	Genève 23	CERN, Route de Meyrin

(continued)

(continued)

Name	citizenship	Country	code	City	address
Dr. Mohamed Hamal	Ma	**Morocco**	60 000	Oujda	Lot Militaire F6/46 Oued Nachef
M. Mahmoud Hosini	Eg	**Egypt**	13762	Kaliubeya	Hospitals, Zagazig, Arab Johayna, Shebeen El Kanater
Ms Julia Hristova	Bg	**Bulgaria**	1692	Sofia	Ovcha kupel 1, bl. 501, vh. G, et. 3, ap.6
Dr. Maksym Kanzyuba	Ua	**Ukraina**	8304	Donetsk	Sc. Research Inst. For Traumatology & Orthopedics, 106, Artema
Dr. P. Kukolowicz*	Pl	**Poland**	25-734	Kielce	Med. PhyS; Dept. Holycross Cancer Centre
Dr. Yves LEMOIGNE	Fr	**France**	74160	Archamps	ESI, Batiment Le Saleve, Site d'Archamps
Dr. C. Lewis*	UK	**UK**	SE5 9RS	London	King's College Hospital, Denmark Hill
Ms Giorgia Loreti	It	**Italy**	40125	Bologna	University of Bologna, Via Broccaindosso 25
Prof. Jim Malone*	Irl	**Ireland**		Dublin 8	Trinity College, St James Hospital
Ms Daniela Neamtu	Ro	**Romania**	200526	Craiova	Calea Bucuresti-Ibfratirii Nr2, 10 E, 12
Dr. P. Peschke*	De	**Germany**	69120	Heidelberg	DKFZ, Im Neuenheim Feld 280
Dr. R. Padovani	It	**Italy**	33100	Udine	SO di Fisica Sanitaria, Ospedale S.Maria della Misericordia
Ms Anna Petrova Nikolova	Bg	**Bulgaria**	1330	Sofia	Serdika, bl. 7, entrance 2, app. 38, floor 6
M. Ohav Rahimian	IL	**Israel**	84801	Beer Sheva	Ben-Gurion Univerity, Jerusalem St. 103/50
Ms Rozalinda Rashkova	Bg	**Bulgaria**	7012	Ruse	Borisiva str., 99, B

(continued)

(continued)

Name	citizenship	Country	code	City	address
Dr. M. Rehani*	India	**Austria**	A-1400	Vienna	IAEA, P.O. Box 100, Wagramer Strasse 5
Dr. Abdeslem Rrhioua	Ma	**Morocco**	60 000	Oujda	Mohammed 1er University, LPTPM, Faculté des Sciences
M. Mohamed Saleh Youssef	Eg	**Egypt**		Cairo	26 Alashgar St. - Dokki
Dr. N. Samet*	Md	**Moldova**	2028	Chisinau	SoS Hancesti 20, App 130
M. Walaa Seif Hamouda	Eg	**Egypt**		Cairo	Cairo-Giza University, Faculty of Science, Physics department
Ms Zdravka Spasova	Bg	**Bulgaria**	1836	Sofia	j.k."Levski" V, bl.4, vh. B,et. 1, ap.21
Dr. M.A. Staniszewska	Pl	**Poland**	90-419	Lodz	Med. Imaging Dept. Medical University, Al Kosciuski, Poland
Dr. Davide Vité	It	**France**	74160	Archamps	ESI, Batiment Le Saleve, Site d'Archamps

Lecturers' names are in **bold**

The Class Room

Lunches taken together

Exercises in the Computing Room

PC Exercises with Renato Padovani

At CERN Radiation Protection laboratory

Discussion with a Radioprotection officer

Supervisor Stelios Christofides (CY)

Medan Rehani (IAEA, Vienna)

Cornelius Lewis (Kings' College, UK)

Nina Samet (Oncology Institute, MD)

Free Discussion Round Table

Farewell Dinner organized by students

Part I
General Radiation Protection Principles

International Safety Standards

Madan Rehani

Abstract International Radiation Safety Standards are developed by the IAEA under its mandate, in cooperation with a number of international organizations and based on scientific data provided by the United Nations Scientific Committee on Effects of Atomic Radiation (UNSCEAR) and radiation protection recommendation by the International Commission on Radiological Protection (ICRP). The International Basic Safety Standard (BSS) published in 1996 is the current standard. National authorities translate this into their national regulations. What is mandatory in countries are the national regulations. In Europe the European BSS is applied. Both International and European BSS are undergoing revision at this moment. This presentation will cover mechanism by which Standards are developed, salient features of the BSS and what changes are coming up. For example, some are: requirements for licensees to have medical physicists in different areas, in particular where higher radiation doses are imparted to patients; there are clarifications on the role of medical physicists; the use of Diagnostic Reference Levels (DRLs) is further strengthened; clarifications are being introduced to avoid low image quality; the concept of radiological audit has been introduced and the need for patient dose recording is emphasized.

1 Introduction

International Radiation Safety Standards form a family of Standards developed by the IAEA in cooperation with large number of international bodies and Member States. The Standards that are relevant to radiation safety in medical application of radiation are International Basic Safety Standards (commonly called as BSS). The BSS was last published in 1996 [1] and is currently undergoing revision. The BSS

M. Rehani (✉)
International Atomic Energy Agency, Vienna, Austria
e-mail: M.Rehani@iaea.org

Y. Lemoigne and A. Caner (eds.), *Radiation Protection in Medical Physics*,
NATO Science for Peace and Security Series B: Physics and Biophysics 1,
DOI 10.1007/978-94-007-0247-9_1,

provides requirements for a legal framework that countries can adopt in for their regulations and legislation. The requirements established in BSS are governed by the objectives, concepts and principles of the Fundamental Safety Principles [2] for ensuring protection of people and the environment. Current revision process draw upon information derived from experiences of the Member States in applying the requirements of the BSS, and from experience in many countries in the use of radiation and nuclear techniques. It also draws upon extensive research and development work by national and international scientific and engineering organizations on the health effects of radiation and on techniques for the safe design and operation of sources.

Why International Standard? In the absence of international Standrds, different countries will have different requirements. For example, a country A may have a dose limit of 20 mSv/year and another country 50 mSv/year, yet another 100 mSv/year and also another 10 mSv/year. Obviously the result will be confusion. Another example, one country may allow 1 month leave for all those occupationally exposed to radiation whereas others not. Requirements for putting personal monitoring badge over the lead apron may be in some country and in another under the lead apron.

Figure 1 shows the organizations in cosmic scheme. Although the International Atomic Energy Agency (IAEA) takes the lead in coordinating development of BSS, there is involvement of United Nations Scientific Committee on Effects of Atomic Radiation (UNSCEAR) and International Commission on Radiological Protection (ICRP), besides number of regional organizations, international organizations and ultimately more than 140 Member States of the IAEA.

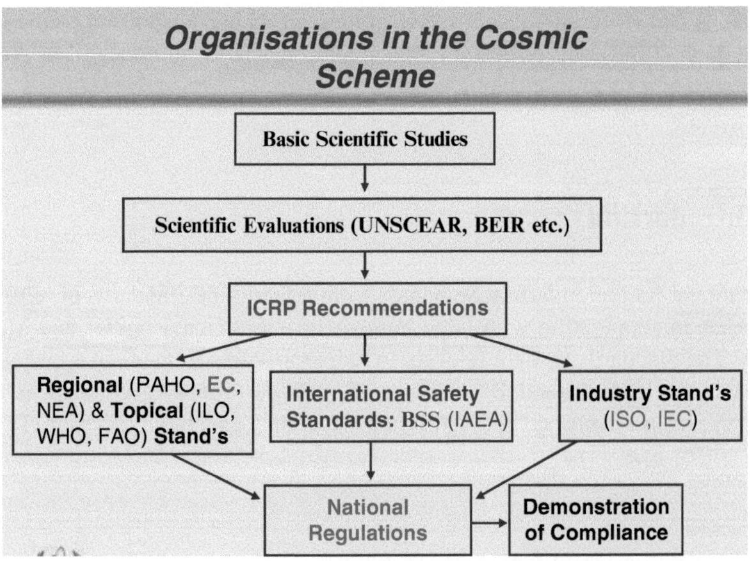

Fig. 1 The organizations in cosmic scheme in framing radiation safety standards and regulations

In addition to international BSS, there is European BSS that is prepared under EURATOM and following Directives: 96/29/EURATOM, 97/43/EURATOM and Medical Device Directive.

The natural question that is normally asked is: Is BSS mandatory? The answer is No. Then what is legally applicable? It is the national regulations. The role of International Standards is that they are robust and sound based on very wide consensus and thus most countries can adopt them. Also where national regulations are lacking, international standards provide acceptable system for legal authorities. The last report on medical exposure was released in 2000 and the next one is going to be released in 2009 itself.

ICRP is an independent charity. It was established in 1928 to advance for the public benefit the science of Radiological Protection, in particular by providing recommendations and guidance on all aspects of protection against ionizing radiation. It offers its recommendations to regulatory and advisory agencies and provides advice intended to be of help to management and professional staff with responsibilities for radiological protection. While ICRP has no formal power to impose its proposals on anyone, in fact legislation in most countries adheres closely to ICRP recommendations.

Among the important aspects of the ICRP are providing protection philosophy and also the dose limits for occupational and public exposure. Initially the basis for dose limit for occupational exposures was avoidance of deterministic harm and for that reason in 1934, a dose limit of ~500 mSv was recommended being erythema dose. The current recommendation of 20 mSv/year was given in ICRP Publication 60 in 1990 that covers stochastic risks. In its recent recommendation in ICRP Publication 103, the tissue weighting factor for breast has been increased from 0.05 to 0.12 (by 140%) and for gonads decreased from 0.2 to 0.08, that is, by $\approx$60% [3].

UNSCEAR was established by the General Assembly of the United Nations in 1955. Its mandate in the United Nations system is to assess and report levels and effects of exposure to ionizing radiation. Governments and organizations throughout the world rely on the Committee's estimates as the scientific basis for evaluating radiation risk and for establishing protective measures.

The IAEA is an independent international organization under UN family. The Standards as developed by the IAEA are based on the following principles of radiation protection and safety in the **Fundamental Safety Principles** [2]:

Safety Principle 1: The prime responsibility for safety must rest with the person or organization responsible for facilities and activities2 that give rise to radiation risks.

Safety Principle 2: An effective legal and governmental framework for safety, including an independent regulatory body, must be established and sustained.

Safety Principle 3: Effective leadership and management for safety must be established and sustained in organizations concerned with, and facilities and activities that give rise to, radiation risks.

Safety Principle 4: Facilities and activities that give rise to radiation risks must yield an overall benefit.
Safety Principle 5: Protection must be optimized to provide the highest level of safety that can reasonably be achieved.
Safety Principle 6: Measures for controlling radiation risks must ensure that no individual bears an unacceptable risk of harm.
Safety Principle 7: People and the environment, present and future, must be protected against radiation risks.
Safety Principle 8: All practical efforts must be made to prevent and mitigate nuclear or radiation accidents.
Safety Principle 9: Arrangements must be made for emergency preparedness and response for nuclear or radiation incidents.
Safety Principle 10: Protective actions to reduce existing or unregulated radiation risks must be justified and optimized.

BSS require main responsibilities to:

- Registrants and licensees
- Employers

Subsidiary responsibilities include:

- Suppliers
- Workers
- Radiation protection officers
- Medical practitioners & health professionals
- Qualified experts, ethical review committees

The employers, registrants and licencees must ensure that workers not employed by them, but who are engaged in work that involves one of their sources, are properly protected.

1.1 Special Compensatory Arrangements

As per BSS, the conditions of service of workers shall be independent of the existence or the possibility of occupational exposure. Special compensatory arrangements or preferential treatment with respect to salary or special insurance coverage, working hours, length of vacation, additional holidays or retirement benefits shall neither be granted nor be used as substitutes for the provision of proper protection and safety measures to ensure compliance with the requirements of the Standards.

For pregnant workers, BSS state that a female worker should, on becoming aware that she is pregnant, notify the employer in order that her working conditions may be modified if necessary. The notification of pregnancy shall not be considered a reason to exclude a female worker from work; however, the employer of a female worker who has notified pregnancy shall adapt the working conditions in respect of occupational exposure so as to ensure that the embryo or fetus is afforded the same broad level of protection as required for members of the public. Employers shall make every reasonable effort to provide workers with suitable alternative employment in circumstances where it has been determined, either by the Regulatory Authority or in the framework of the health surveillance program required by the Standards, that the worker, for health reasons, may no longer continue in employment involving occupational exposure.

In addition to requirements for justification and optimization, there are requirements for **Guidance levels (or diagnostic reference levels)** for medical exposure. The guidance levels are intended:

(a) To be a reasonable indication of doses for average sized patients
(b) To be established by relevant professional bodies in consultation with the Regulatory Authority following the detailed requirements of Appendix II in BSS and the guidance levels given in Schedule III
(c) To provide guidance on what is achievable with current good practice rather than on what should be considered optimum performance
(d) To be applied with flexibility to allow higher exposures if these are indicated by sound clinical judgment and
(e) To be revised as technology and techniques improve

References

1. Food and Agriculture Organization of The United Nations, International Atomic Energy Agency, International Labour Organization, OECD Nuclear Energy Agency, Pan American Health Organization, World Health Organization, International Basic Safety Standards for Protection against Ionizing Radiation and for the Safety of Radiation Sources, Safety Series No.115, IAEA, Vienna (1996).
2. European Atomic Energy Community, Food and Agriculture Organization of The United Nations, International Atomic Energy Agency, International Labour Organization, International Maritime Organization, OECD Nuclear Energy Agency, Pan American Health Organization, United Nations Environment Programme, World Health Organization, Fundamental Safety Principles; Safety Fundamentals, IAEA Safety Standards Series No. SF-1, IAEA, Vienna (2006).
3. International Commission on Radiological Protection, ICRP Publication 103: Recommendations of the ICRP, Annals of the ICRP Volume 37/2–4, Elsevier (2007).

Radiation Units, Limits and Dose Constraints

Madan Rehani

Abstract A number of radiation units exist for representing radiation dose to patients, staff, public and quantities for radiation equipment performance in radiological practice. What is important is how ably to communicate these to medical professionals. The concept of air kerma, a measurable quantity outside the human body has been recommended by the IAEA and ICRU (International Commission on Radiation Units and Measurements). From air kerma, one estimates the entrance air kerma on patient or other quantities such as $CTDI_w$ or $CTDI_v$. Also the kerma area product (KAP) is useful quantity that is easily measured in particular in fluoroscopic procedures. The organ doses have direct relationship with biological effects and can be estimated from entrance air kerma. By using the tissue weighting factor, one estimates the dose equivalent and effective dose. In specific situation of interventional procedures, cumulative air kerma at interventional reference point has been described. Similarly there are quantities for staff dose estimation based on measurable values and estimation of dose equivalent and effective dose. There is a concept of dose constraint that is applied to carers and comforters of patients. This concept is also applied in occupational protection. Radiation dose limits are given for staff and members of the public, dose constraints for comforters and there are no dose limits for patients but the concept of diagnostic reference level applies. The lecture will cover these quantities, explain the role played by international organizations and discuss how they can be used in day-to-day practice in hospitals.

1 Introduction

Radiation exposures resulting from radiological procedures constitute the largest part of the population exposure from artificial radiation. There are a number of dose quantities for staff, patient, carers & comforters and for members of the public.

M. Rehani (✉)
International Atomic Energy Agency, Vienna, Austria
e-mail: M.Rehani@iaea.org

Y. Lemoigne and A. Caner (eds.), *Radiation Protection in Medical Physics*,
NATO Science for Peace and Security Series B: Physics and Biophysics 1,
DOI 10.1007/978-94-007-0247-9_2,

Moreover, the needs vary in different situations. For example, in interventional procedures in adults, the focus is on avoidance of deterministic injuries whereas in children the reducing the probability of stochastic risk is important. One needs to know the dose to the skin for deterministic effects on skin (erythema) and dose area product for assessing the stochastic effects. Thus the dose quantities vary with situation and there is need to be aware about the information each dose quantity can give and the limitation associated.

One of the important aims of patient dosimetry with respect to X rays used in medical imaging is to determine dosimetric quantities for the establishment and use of guidance levels (diagnostic reference levels, DRL) and for comparative risk assessment. In the latter case, the average dose to the organs and tissues at risk should be assessed. An additional objective of dosimetry is the assessment of equipment performance as a part of the quality assurance process. Although it is of interest to make measurements directly on the patient, this is something not practicable in most situations. Therefore measurements using a standard phantom to simulate the patient are done for the control of technical parameters, for the comparison of different systems and for optimization.

The absorbed dose to tissue is important in radiotherapy whereas in nuclear medicine, organ doses and effective dose is useful.

2 Definitions and Interpretations

Absorbed dose is the energy absorbed per unit mass at a given point. The unit is the joule per kilogram (J kg^{-1}) and is given the special name gray (Gy).

Organ dose is a quantity defined in ICRP Publication 60 in relation to the probability of stochastic effects (mainly cancer induction) as the absorbed dose averaged over an organ, i.e., the quotient of the total energy imparted to the organ and the total mass of the organ. The unit is the joule per kilogram and is given the special name gray (Gy).

Equivalent dose to an organ or tissue is the organ dose corrected by a radiation weighting factor that takes account of the relative biological effectiveness of the incident radiation in producing stochastic effects. This correction factor is numerically 1 for X rays. The unit is the joule per kilogram (J kg^{-1}) and is given the special name Sievert (Sv).

Effective dose is a quantity defined in ICRP Publication 60 as a weighted sum of equivalent doses to all relevant tissues and organ with the purpose "to indicate the combination of different doses to several different tissues in a way that is likely to correlate well with the total of the stochastic effects". This is, therefore, applicable even if the absorbed dose distribution over the human body is not homogeneous. The unit is the joule per kilogram (J kg^{-1}) and is given the special name Sievert (Sv).

The use of effective dose for patients has to be done with caution, as indicated in the UNSCEAR 2000 report to the UN, "effective dose should not be used directly for estimating detriment from medical exposure … by application of the nominal fatality probability coefficients. Such assessments would be inappropriate and serve no purpose in view of the uncertainties arising from potential demographic differences (in terms of health status, age and sex), between particular population of patients and those from general populations for whom ICRP derived the risk coefficients ... effective dose could broadly underestimate the detriment from diagnostic exposures of young patients by a factor of 2 and, conversely, could overestimate the detriment from old patients by a factor of at least 5. ... Notwithstanding the above caveat ... practice in diagnostic radiology is summarized for comparative purpose, principally in terms of effective dose to the exposed individuals ... taking into account the number of procedures, collective effective dose over exposed populations".

It is possible, therefore, to use effective dose and even collective dose for medical diagnostic exposure as long as this is done only for comparative purposes and for the same or similar patient populations, and it would require additional considerations or significant corrections if we try to use them to compare with other populations.

Air kerma in air is the sum of kinetic energy of all charged particles liberated per unit mass. A number of publications in the past have expressed measurements in terms of absorbed dose to air. Recent publications and a soon-to-be-published IAEA Code of Practice point out the experimental difficulty in determining the dose to air, especially in the vicinity of an interface, and that, in reality, what the dosimetry equipment registers is not the energy absorbed from the radiation by the air, but the energy transferred by the radiation to the charged particles resulting from the ionization. For these reasons the IAEA Code of Practice and ICRU Report 74 recommend the use of air kerma rather than absorbed dose to air. The unit is the joule per kilogram ($J\ kg^{-1}$) and is given the special name gray (Gy).

This correction applies to the quantities determined in air, such as entrance surface air kerma (rather than entrance surface air dose), computed tomography air kerma index (instead of computed tomography dose index), kerma area product (rather than dose area product) and air kerma area length (rather than dose length product (DLP)).

The above recommendation refers to air. When referring to tissues, it is also correct to estimate absorbed dose to the skin, by applying the necessary correction coefficient to obtain the absorbed dose to the tissue from the air kerma.

Definitions may not always be important. For example, it may be extremely difficult for most people and professionals to define temperature, pressure, length. But everyone uses these and can have a "feel" of these quantities. Thus definitions without feel are meaningless. To get a feel, let us see following:

How much radiation do we get from natural sources? One may say: around 1–3 mSv.
Which dose quantity is this? It is effective dose.
How much radiation a patient gets in chest radiograph? Typically around 0.02 mSv.
Again what dose quantity is this? Effective dose.
The dose limit for staff for extremities is 500 mSv. This is equivalent dose.

Entrance surface air kerma is useful in radiography and mammography when it comes to diagnostic procedure of relating to equipment radiation for quality assurance purpose. But specific quantities are needed such mean glandular dose (MGD) for breast tissue. Similarly kerma area product (KAP or DAP is useful for assessing the stochastic risk to patients. Specfic quantities in computed tomography are CTDI (Computerized Tomography Dose Index).

2.1 Specific Dose Quantities in Computerized Tomography

There are specific dose descriptors in CT which are discussed in this part and these are: (1) Computerized Tomography Dose Index (CTDI), (2) Dose Length Product (DLP) and (3) Effective dose (E). It should be noted that the International Commission on Radiation Units and Measurements (ICRU) has recently recommended the use of the quantity CT air kerma index for CT.[(69)] However, since the audience of this article being primarily radiologists rather than medical physicists, the authors decided to use the term CTDI for simplicity in understanding till the newer term becomes familiar.

2.1.1 Computerized Tomography Dose Index

CTDI integrates the radiation dose imparted within and beyond a single slice and it is defined by the following equation:

$$CTDI = \frac{1}{T}\int_{-\infty}^{+\infty} D(z)dz \qquad (1)$$

T is the nominal slice thickness

D(z) is the dose profile along a line parallel to the Z-axis (tube rotation axis)

For CTDI measurement, two polymethylmethacrylate (PMMA) cylinders of 14 cm length are used. For head examinations, a phantom diameter of 16 cm is used and for body, a phantom diameter of 32 cm is applied. The phantoms are called, respectively, as the head and body CTDI phantoms. CTDI is measured using a specially designed pencil ionization chamber with an active length of 100 mm both in free air at the centre of rotation ($CTDI_{air}$) and within the holes of the two phantoms. $CTDI_c$ and $CTDI_p$ are defined respectively as the CTDI values measured with a pencil chamber dosimeter positioned in the centre and in the periphery of the PMMA head or body phantom.

- $CTDI_w$ is used for approximating the average dose over a single slice in order to account for variations in dose values between the center and the periphery of the slice. It is defined by the following equation:

$$CTDI_w = \frac{1}{3}CTDI_c + \frac{2}{3}CTDI_p \qquad (2)$$

- $CTDI_p$ is the average of the four $CTDI_p$ values measured in the periphery of the phantom (12, 3, 6 and 9 o' clock).
- $CTDI_{vol}$ represents the radiation dose in one tube rotation in MDCT and allows for variations in exposure in the z direction when the pitch (p) (pitch is the ratio of table feed in one rotation (I) to slice collimation (NT)).

$$CTDI_{vol} = NT / I * CTDIw \quad (3)$$

$$\mathrm{CTDI}_{\mathrm{vol}} = \mathrm{CTDI}_{\mathrm{w}} / \mathrm{p} \quad (4)$$

This equation applies when p is not equal to 1.

CTDI is measured in mGy and the display of CTDI value on the CT console is strongly recommended.[70] It should be noted that CTDI has a number of limitations. It is measured by using a standardized, homogeneous, cylindrical phantom and therefore it possibly differs from the dose for objects of substantially different size, shape, or attenuation, like the human body.[71] It is expressed as dose to air, not dose to tissue and it is not sufficient for slice collimations greater than 10 cm such as those of 256 or 320 CT scanners. Finally, it does not indicate the dose to a specific point in the scan volume when the patient table remains stationary for multiple scans, such as for interventional or perfusion CT.

2.1.2 Dose Length Product

DLP is used to calculate the dose for a series of slices or a complete examination and is defined by the following equation:

$$DLP = \sum_{i}^{N} CTDI_w TN$$

i represents each one of the individual N scans of the examination that covers a length T of patient anatomy. It is a way to evaluate the total radiation dose given to the patient during a specific examination. This practically means that for a given technical protocol with certain $CTDI_{vol}$, the DLP of two scanning regions with different lengths will be different.

3 Occupation Doses

In monitoring occupational exposures to external radiation, individual dosimeters measure the personal dose equivalent HP(10). This measured value is taken as an assessment of the effective dose under the assumption of a uniform whole body exposure. For internal exposure, committed effective doses are generally determined from an assessment of the intakes of radionuclides from bioassay measurements or other quantities (e.g., activity retained in the body or in daily excreta). The radiation dose is determined from the intake using recommended dose coefficients.

The doses obtained from the assessment of occupational exposures from external radiation and from intakes of radionuclides are combined for the assignment of the value of total effective dose, E, for demonstrating compliance with dose limits and constraints

4 Dose Constraints

A prospective and source-related restriction on the individual dose from a source, which provides a basic level of protection for the most highly exposed individuals from a source, and serves as an upper bound on the dose in optimization of protection for that source.

For occupational exposures, the dose constraint is a value of individual dose used to limit the range of options considered in the process of optimization.

For public exposure, the dose constraint is an upper bound on the annual doses that members of the public should receive from the planned operation of any controlled source.

Part II
Radiobiology Principles

Molecular Aspects of Radiation Biology

Peter Peschke and Günther H. Hartmann

1 Introduction

The application of molecular biology to radiobiology and radiation oncology has considerably improved our understanding of cellular radiation responses of both tumor and normal cells. With the current rapid developments in molecular and cellular biology a clear need exists to integrate this knowledge into basic and preclinical research in radiation oncology. Consequently, molecular radiation biology/oncology will thus provide tools to develop new strategies of individualized therapy and molecular targeting in modern radiation oncology, which eventually will result in the improvement of tumor responses and a reduction of normal tissue reactions.

2 The Basic Interaction Processes

Radiation applied to biological systems results in a transfer of energy from radiation to the system. Accordingly, the dose of an ionizing radiation is expressed by the (radiation) energy absorbed per unit mass. There two main categories of energy transfer to biological systems on a molecular level:

- Via direct effects (about one-third): direct effects encompass all the physical interactions between the radiation and the atoms of the biological system resulting in the ionization or excitation of the atoms, where almost any physical interaction leads to the production of energetic secondary electrons and associated loss/transfer of their energy
- Via indirect effects (about two-third): indirect effects encompass all the physical/chemical interactions of radicals produced by radiation in biological systems; the pathways of radicals produced in water are a good model to understand such

P. Peschke (✉) and G.H. Hartmann
Department of Radiation Oncology, German Cancer Research Center, Research Program Imaging and Radiooncology, Im Neuenheimer Feld 280, D-69120 Heidelberg, Germany
e-mail: p.peschke@dkfz-heidelberg.de

Y. Lemoigne and A. Caner (eds.), *Radiation Protection in Medical Physics*, NATO Science for Peace and Security Series B: Physics and Biophysics 1,
DOI 10.1007/978-94-007-0247-9_3,

processes, however, that occurring in biological system are important as well. An overview over the main processes in water is given below:

Hydroxyl radical:

$$H_2O + H_2O \rightarrow H_3O^+ + OH^-$$

Solvated electrons:

$$e^- + [H_2O^+] \rightarrow e^-_{aqu.}$$

Hydroxyl radical:

$$e^-_{aqu.} + H_2O \rightarrow OH^-$$

Hydrogen peroxide:

$$OH^- + OH^- \rightarrow H_2O_2$$

3 Effects at DNA Level

The cell nucleus, a membrane-enclosed organelle found in eukaryotic cells, contains most of the cell's genetic material, organized as multiple long linear molecules of deoxyribonucleic acid (DNA) with a large variety of proteins, such as histones, to form chromosomes.

The DNA contains the genetic instructions used in the development and functioning of all known living organisms and some viruses. A schematic drawing of the structures forming a nucleus cell down to the DNA molecule is shown in Fig. 1.

Various damages of the DNA molecule can occur via the direct as well as the via the indirect effect pathways. A summary is given in Fig. 2.

4 Processes Following DNA Damage

Once a DNA lesion of any type is produced, two main categories of processes are invoked (Fig. 3): (a) processes that "handle" the DNA lesion such that function and/or survival of the cell are maintained (b) processes that ultimately lead to malignant transformation or cell death.

4.1 Repair Processes

Because DNA is the repository of genetic information in each living cell, its integrity and stability are essential to life. Therefore, DNA repair processes exist in both prokaryotic and eukaryotic organisms, and many of the proteins involved have been

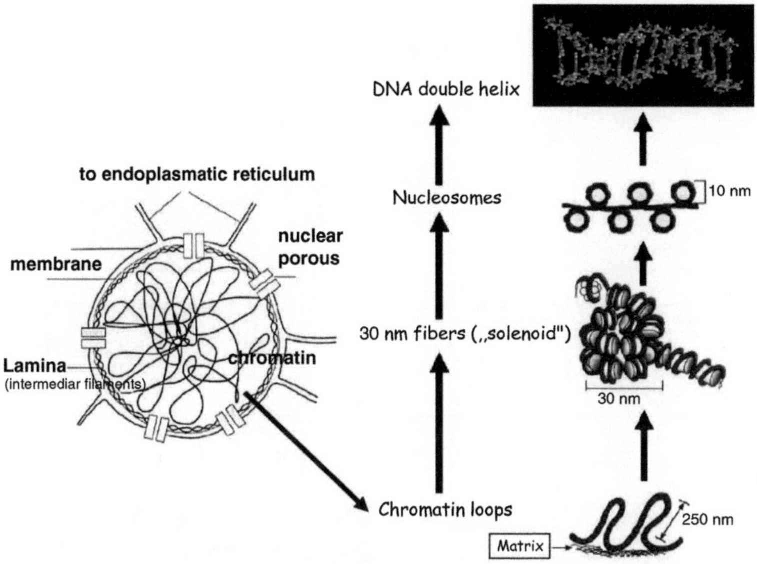

Fig. 1 Schematic drawing of the structures forming a nucleus cell down to the DNA molecule

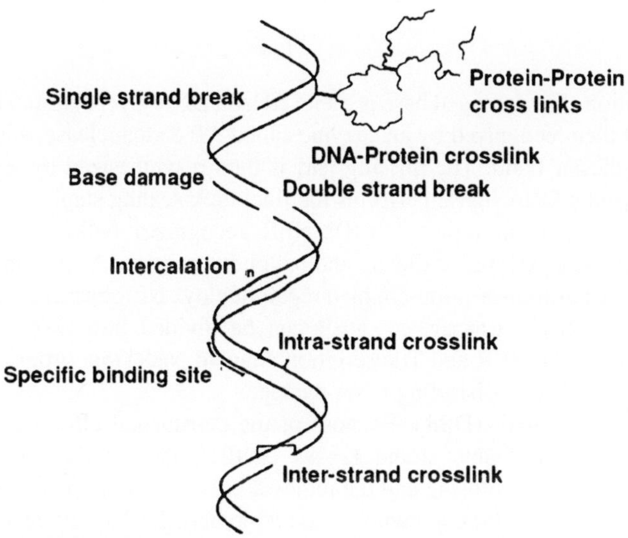

Fig. 2 DNA lesions following radiation

highly conserved throughout evolution. With respect to radiation effects in biological systems, the different repair processes are of particular interest and have been extensively investigated since a long time. Briefly, one can differentiate between:

- Base excision repair (BER): BER (active throughout the cell cycle) (BER), repairs damage to a single base caused by oxidation, alkylation, hydrolysis, or

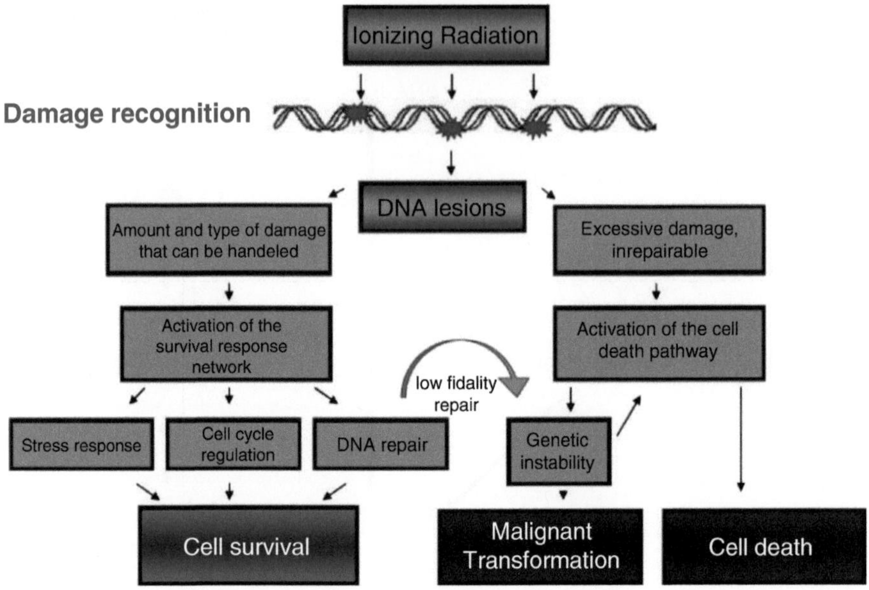

Fig. 3 Possible processes following DNA damage

deamination. The damaged base is removed by a DNA glycosylase. The "missing tooth" is then recognized by an enzyme called AP endonuclease, which cuts the Phosphodiester bond. The missing part is then resynthesized by a DNA polymerase, and a DNA ligase performs the final nick-sealing step.

- Nucleotide excision repair (NER): NER recognizes bulky, helix-distorting lesions such as pyrimidine dimers and 6,4 photoproducts. A specialized form of NER known as transcription-coupled repair deploys NER enzymes to genes that are being actively transcribed. NER can be divided into two sub-pathways (Global genomic NER and Transcription coupled NER) that differ only in their recognition of helix-distorting DNA damage.
- Double strand repair (DSR): Because of the detrimental effects of unrepaired and misrepaired double strand breaks (DSBs), the cell devotes significant resources to the monitoring and the removal of such lesions. Three mechanisms exist to repair DSBs: (a) non-homologous end-joining (NHEJ), (b) microhomology-mediated end-joining (MMEJ) and (c) homologous recombination.

Important to know:

DSBs can lead to chromosomal aberrations. Typical findings are shown in Fig. 4.

In summary, the outcome of a repair process ultimately can lead to:

- Accurate repair: The cell survives without mutations.
- Misrepair: The cell survives but at the cost of genetic changes which may include carcinogenesis.

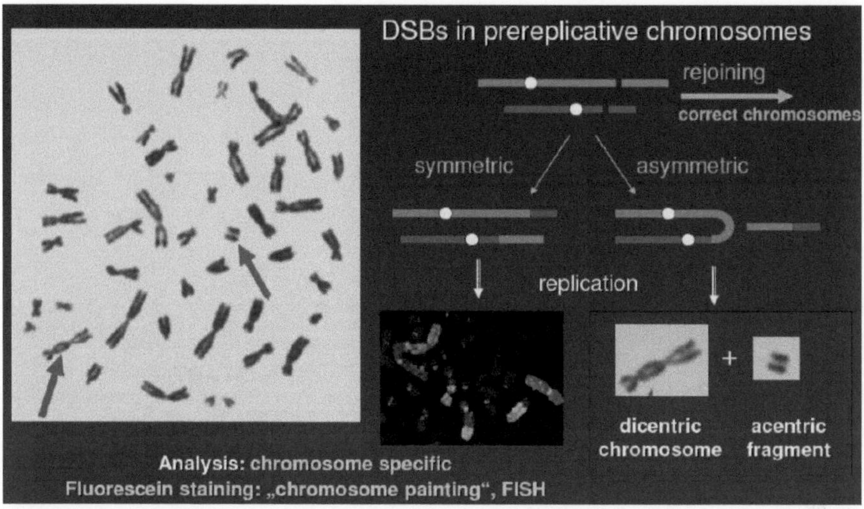

Fig. 4 Chromosomal aberrations after mis-repair of double strand breaks. See color picture in Appendix 1

- Inadequate repair: Inadequate repair leads to cell inactivation or cell death due to (a) mitotic death, (b) apoptosis, or (c) permanent arrest.

4.2 Radiation Induced Processes Influencing the Cell Cycle

The cell cycle consists of four distinct phases as shown in Fig. 5: G1 phase, S phase (synthesis), G2 phase (collectively known as interphase) and M phase (mitosis). The M phase is itself composed of two tightly coupled processes: mitosis, in which the cells chromosomes are divided between the two daughter cells, and cytokinesis, in which the cells cytoplasm divides in half forming distinct cells. Activation of each phase is dependent on the proper progression and completion of the previous one. Cells that have temporarily or reversibly stopped dividing are said to have entered a state of quiescence, called G0 phase.

In the context of radiation damage, cell cycle checkpoints play an important role. They are the control mechanisms that ensure the fidelity of cell division in eukaryotic cells.

- G1 Checkpoint: This checkpoint is located at the end of the cell cycle's G1 phase, just before entry in the S phase, making the key decision of whether the cell should divide, delay division, or enter a resting stage. The G1 checkpoint is where eukaryotes typically arrest the cell cycle if environmental conditions make cell division impossible or if the cell passes into G0 for an extended period.
- G2 Checkpoint: The second checkpoint is located at the end of G2 phase, triggering the start of the M phase (mitosis). In order for this checkpoint to be passed, the cell has to check a number of factors to ensure the cell is ready for mitosis.

Typical radiation-induced effects on these checkpoints are shown in Fig. 6.

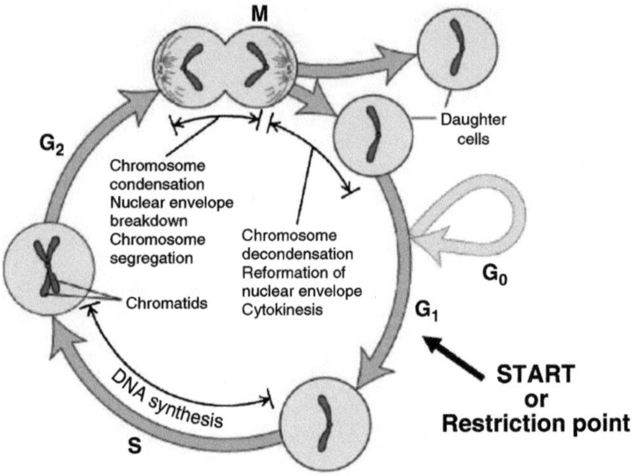

Fig. 5 The intact cell cycle

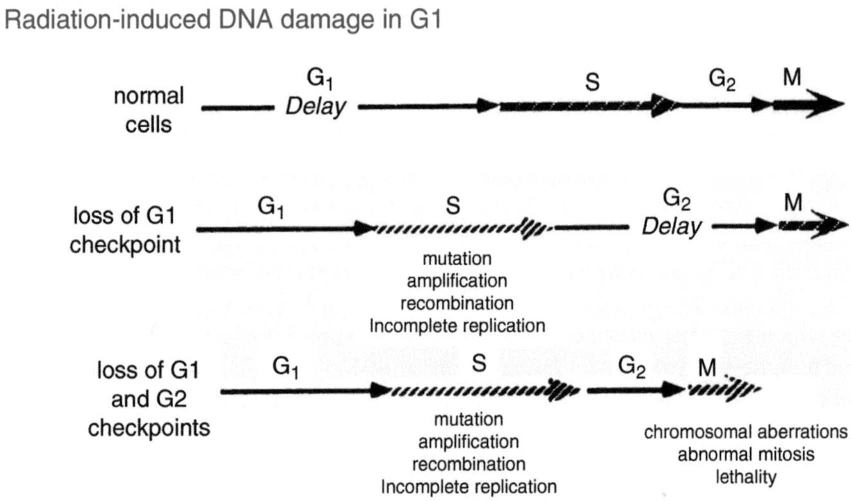

Fig. 6 Radiation-induced effects on the cell cycle

4.3 Other Effects Contributing to Radiation Sensitivity

With respect to the action of radiation-induced radicals in biological systems, oxygen plays an important role. Oxygen significantly contributes to forming DNA-damaging free radicals thus increasing the effectiveness of a given dose of

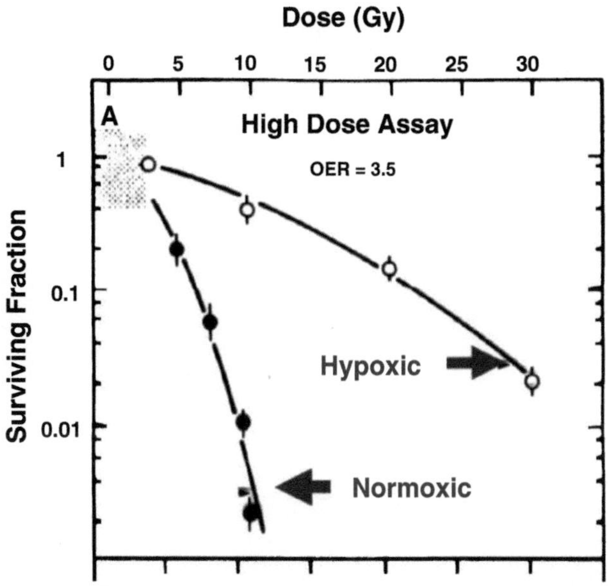

Fig. 7 Difference in radiation sensitivity as obtained in cell survival experiments under normoxic and hypoxic conditions

radiation. Therefore, it is a potent radiosensitizer. Tumor cells in a hypoxic (reduced oxygen) environment may be as much as two to three times more resistant to radiation damage than those in a normal oxygen environment (see Fig. 7).

Much research has been devoted to overcoming this problem including the use of high pressure oxygen tanks, blood substitutes that carry increased oxygen, hypoxic cell radiosensitizers such as misonidazole and metronidazole, and hypoxic cytotoxins, such as tirapazamine.

There is also interest in the fact that high linear energy transfer particles such as ions or neutrons may have an antitumor effect which is less dependent of tumor oxygen because these particles act mostly via direct damage.

A further known substance influencing the induction of radiation effects at the molecular level is glutathione. It is a tripeptide containg a sulfhydrol group and an unusual peptide linkage between the amine group of cysteine and the carboxyl group of the glutamate side chain. Glutathione acts as an oxidative buffer and plays a key role in detoxification by interacting with hydrogen and organic peroxides.

References

1. E.J. Hall, Radiobiology for the Radiologist, Lippincott, Philadelphia, PA (2000).
2. A.H.W. Nias, An Introduction to Radiobiology, Wiley, New York (1998).
3. G.G. Steel, Basic Clinical Radiobiology, Arnold, London (2002).
4. Radiation Damage in DNA – Structure/Function Relationships at Early Times. (Eds: A. F. Fuciarelli and J. D. Zimbrick) Batelle Press, Columbus, OH (1995).

Radiobiology in Radiotherapy

Pawel Kukolowicz

1 Introduction

Very soon after discovery of X-Rays by Wilhelm Conrad Roentgen the new radiation was used for therapeutic purposes. In 1896 several patients were treated in Germany, Austria and France with external beams. A few years later Radium discovered by Maria Skolodowska-Curie and Piere Curie was applied for treatment. The first article on radiotherapy was published in 1896 by French physician Victor Despeignes. In 1904, Joseph Belot published a textbook on radiotherapy in which several chapters were devoted to a historical review of teletherapy (!). [Belot. J. Trait de radiotherapie. Paris: G. Steinheil; 1904]. This time the radiation was used in purely empirical way. There was no idea how to measure the amount of radiation delivered to a patient. One should notice that the term "dose" or more correctly "absorbed dose" used today was directly taken from pharmacology. The first attempts of radiotherapy caused a large number of various skin lesions. This type of answer of skin to radiation, very easily noticed by vision, was of course, in the center of interest. Then one may express, that the radiobiology started parallel to application of ionizing radiation to treatment and diagnosis. In 1897 the results of first regular studies concerning effects of radiation on normal and malignant tissues were published. Slowly radiobiology became a separate branch of science, however always with the strong link to radiation therapy.

Today radiation biology has an important impact on clinical radiation therapy by providing a rationale for implementation of new treatment strategies into radiotherapy. Parallel to radiobiological experiments performed with cells and animals in research laboratories, the so called clinical radiobiology has been developed. The clinical radiobiology is focused on the quantitative description of the application of radiotherapy for the treatment of patients suffering from cancer. In this lecture some ideas of quantitative clinical radiobiology are presented. This knowledge is important

P. Kukolowicz (✉)
Medical Physics Department, Holycross Cancer Center, Kielce, Poland
e-mail: pk@message.pl

Y. Lemoigne and A. Caner (eds.), *Radiation Protection in Medical Physics*,
NATO Science for Peace and Security Series B: Physics and Biophysics 1,
DOI 10.1007/978-94-007-0247-9_4,

for medical physicists because, as we will be informed, the requirement concerning the precision of dose delivery is a result of the very sharp dependence of the treatment outcome on the precision of the dose delivered. Also because in clinical practice more and more often mathematical models are used, that require some mathematical background and this is the role of physicists involved in radiation treatment.

2 Cell Survival Curves

The ionizing radiation like other toxic agents, e.g. cytotoxic drugs, may be used to kill malignant cells. The eradication of every malignant tumour cells leads to local control of the tumour. Therefore in the case of radiotherapy one of the most important tasks is to understand the mechanism of cell's killing after exposure to ionizing radiation and to describe mathematically the dependence of the survival of cells on the dose. The effectiveness of the radiation is presented in the form of a cell survival curve. This is a plot of the surviving fraction of the cells versus dose. Graphically the survival curves are usually plotted on a logarithmic scale. In Fig. 1, two typical experimentally obtained cell survival curve are shown. The most important difference between these curves is seen at small doses. Curve A is always linear while curve B is continually bending at small doses. The shape of surviving curves at small doses plays an important role in radiotherapy. In conventional fractionation scheme the total dose is delivered with daily dose of 1.8–2.2 Gy, five times a week.

An interesting theory explaining the shape of both cell survival curves was proposed by Curtis. According to his model, called as the lethal, potentially lethal

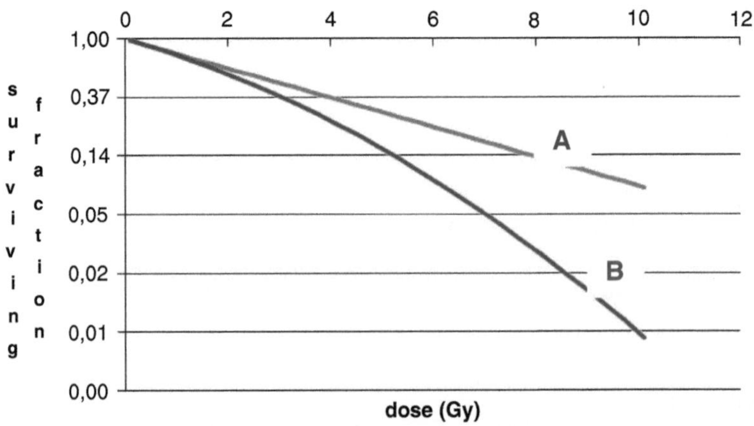

Fig. 1 Surviving curves. (**A**) $SF(d) = \exp(-\alpha \cdot d)$, (**B**) $SF(d) = \exp(-\alpha \cdot d - \beta \cdot d^2)$

damage model, in which the ionizing radiation may produce two different types of lesions: repairable (i.e. potentially lethal) lesions, and non-repairable (i.e. lethal) lesions. Curve A describes a situation in which a cell attacked with radiation is neutralized – this type of lesion cannot be repaired. Such exponential (linear on logarithmic scale) cell survival curves are typical for cells exposed to high LET radiation. Mathematically, this type of curves can be approximated by the formula:

$$SF(d) = \exp(-\alpha \cdot d),$$

where SF is the surviving fraction, d is the dose, α is a parameter describing the sensitivity of cells to radiation for non-repairable lesions.

Curve B is more complex. It describes a situation in which some of the lesions can be repaired. Mathematically, this involves the linear and quadratic terms:

$$SF(d) = \exp(-\alpha \cdot d - \beta \cdot d^2)$$

In this equation the new factor is a parameter describing the sensitivity of cells to radiation for repairable lesions, while the meaning of the other factors remain unchanged.

A very interesting application of the linear-quadratic model has been found in fractionated radiotherapy. We will deal with this application in the last part of this lecture.

One may ask whether the shape of the cell survival curves is always the same regardless of the dose already delivered – in other words thinking of fractionated radiotherapy, the question is: whether the survival fraction is constant irrespective of the session of the treatment. The answer is positive, but only if the gap between fractions is longer than about 8 h.

3 Local Control of a Tumour

It has already been pointed out that the local control of a tumour is achieved when all malignant cells have been killed. Let us assume that the initial number of malignant cells of a tumour is N_0. The average number of cells surviving N_{surv} after dose d, is given by formulae:

$$N_{surv} = N(0) \cdot SF(d) = N(0) \exp(- \alpha \cdot d - \beta \cdot d^2) \quad (1)$$

The process of cell killing is a stochastic one. What is the probability that all cells will die? Mathematically, this question may be formulated in the form of a conditional probability:

$$P(0 | N_{surv}) \quad (2)$$

The answer is given by Poisson statistics:

$$P(0|N_{surv}) = \exp(-N_{surv}) = \exp(-N_0 \cdot \exp(-N_0 \cdot {}^{-\alpha \cdot d - \beta \cdot d^2})) \qquad (3)$$

The tumour control probability (TCP) depends only on the average number of surviving malignant cells. Let us imagine that in a hypothetical sample of 100 tumours *the average* number of surviving cells is one. It means that in all tumours 100 cells will survive, however in reality we do not know the numbers of cells which survive in each tumour. We know the distribution of surviving cells, which is given by Poisson statistics. The Poisson statistics informs us that in 37 tumours all cells will be neutralized and the tumour control will be obtained. Figure 2 shows the dependence of the TCP on dose for a tumour with initial number 10^6 cells, the radiation sensitivity of which is described by $\alpha = 0{,}215\ Gy^{-1}$, and $\beta = 0{,}0215\ Gy^{-2}$.

From clinical experience it is well known that the earlier radiotherapy starts the greater is the chance to cure a patient. Let us imagine a situation where accidentally a very small tumor of 5 mm diameter has been detected. The patient was given a possibility to be treated with irradiation but he refused (unfortunately it sometimes happens). At this stage of the disease there is a chance to control a tumor close to 90%. A few months later the patient changed his mind. Unfortunately, the tumor diameter doubled. Let us estimate the chance to control the tumor. From formulae 1 for initial tumor size the TCP was:

$$0.9_{5mm} = \exp(-N_{5mm} \cdot SF) \qquad (4)$$

Few months later, assuming that the number of malignant cells is proportional to the volume of the tumor, the TCP of this tumor is described as follows:

$$TCP_{10mm} = \exp(-N_{10mm} \cdot SF) = \exp(-8 \cdot N_{5mm} \cdot SF) = (0.9)^3 = 0.35 \qquad (5)$$

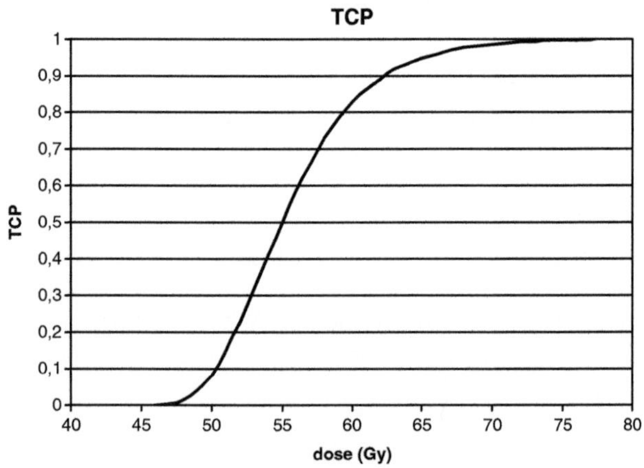

Fig. 2 The dependence of the probability of tumor control on dose

This example illustrates how much the result of radiotherapy depends on an early diagnosis.

There is a second very important factor that strongly influences the result of radiotherapy – the precision of dose delivery. It is so important because the steepness of the dose-response curve is quite high. The steepness of the dose-response curve is very often quantified with the normalized dose-response gradient (NDG). The normalized dose-response gradient is defined by the formulae:

$$\gamma_{D0} = D_0 \cdot \left. \frac{dTCP}{dD} \right|_{D0} \tag{6}$$

The interpretation of the normalized dose-response gradient (NDG) is very simple. Numerically, the NDG describes the increase (decrease) of response in percentage points for a 1% increase (decrease) in dose. It is relatively easy to express analytically the NDG for TCP at a dose for which the TCP= 1/e =0.37. Using formulae 1 we obtained:

$$NDG_{0.37} = \frac{\ln(N_0)}{e} \tag{7}$$

where N_0 is an initial number of malignant cells.

It is estimated that in a 1 cm^3 tumour there is about 10^5 malignant cells to be killed. Then for tumour of 1 cm diameter there are 0.5×10^5 malignant cells. This results in NDG values of about 4%. Several clinical studies showed that the steepness of the dose-response curves are really high but much shallower than predicted by theory. Observed clinical NDG values are smaller than three. This phenomenon is explained by patient-to-patient variability in tumour biological parameters. Radiosensitivity of human tumour cells is different even for the same histopathological tumour types. Deacon and co-workers in 1984 published a paper with the data on surviving fraction at 2 Gy dose for 51 human tumour cell lines which covered 17 different histopathological tumour types. Their results show that the surviving fraction for the same tumours (histopathologically) might differ. Unfortunately, there are no methods to estimate before treatment the actual sensitivity of a tumour. In clinical practice we make decisions based on general information about the tumour. A wide range of different patients are treated as having identical tumours and consequently the same dose is delivered. Some of them are over-treated (too high a dose is delivered), others are under-treated (too small dose is delivered). These results in a less steep dose-response curve, which may be presented graphically in the following way. Let us consider a hypothetical clinical situation. A group of 500 patients were treated with radiation. The initial size of all tumours was the same but the sensitivity of the tumours' cells was different. To make the analysis simpler let us divide the whole group into 5 fully homogenous subgroups and assume that subgroups 1, 2, 3, 4 and 5 to have a TCP of 50% for a total dose of 35, 40, 45, 50 and 55 Gy respectively. Figure 3 shows the dose-response curve for each group.

Table 1 The expected number of cured patients

Dose (Gy)	Group 1	Group 2	Group 3	Group 4	Group 5	Whole group	TCP
30	0.5	0.0	0.0	0.0	0.0	0.5	0.001
35	49.8	1.7	0.0	0.0	0.0	51.5	0.103
40	91.2	50.1	3.5	0.0	0.0	144.8	0.290
45	98.8	88.9	49.9	5.8	0.0	243.5	0.487
50	99.8	98.0	86.6	50.3	8.2	343.0	0.686
55	100.0	99.7	97.1	84.7	50.3	431.7	0.863
60	100.0	99.9	99.4	96.1	82.8	478.1	0.956
65	100.0	100.0	99.9	99.0	94.9	493.8	0.988
70	100.0	100.0	100.0	99.8	98.6	498.3	0.997
75	100.0	100.0	100.0	99.9	99.6	499.5	0.999
80	100.0	100.0	100.0	100.0	99.9	499.9	1.000
85	100.0	100.0	100.0	100.0	100.0	500.0	1.000
90	100.0	100.0	100.0	100.0	100.0	500.0	1.000

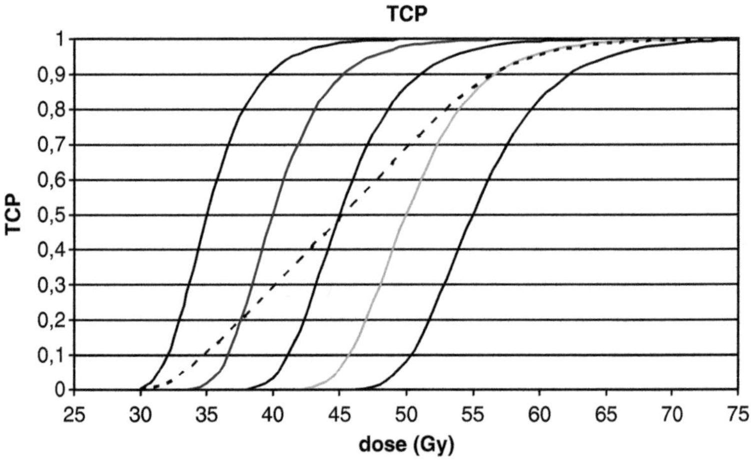

Fig. 3 The TCP curves for each homogenous group of patients separately and for the whole group (*dotted line*). See color picture in Appendix 1

Table 1 presents the expected numbers of cured patients in each group for doses from the Dmin to Dmax. The last column presents the expected number of cured patients for all 500 patients for each dose. According to these results, the resultant tumor-response curve is drawn in Fig. 3 with dotted line. As it has already been pointed out, the steepness of the dose-response curve for the whole group is smaller than it is for each individual homogenous group.

How is the concept of the normalized dose-response gradient can be used in clinical practice? One obvious application of the NDG concept is to use it as a multiplier that converts a relative change in dose into a change in response. For example, the increase of the total dose from 64 to 66 Gy in a schedule with a fraction dose of 2 Gy increases the local control by about:

$$\Delta TCP = 2 \cdot \frac{66Gy - 64Gy}{G4Gy} \cdot 100\% \approx 6\%$$

In this example it was assumed that the NDG equals to two. Always when we calculate the advantage from the dose escalation, we should remember that some higher doses are also delivered to the healthy structures which may result in a higher risk of healthy tissue damage. In order to increase the tumour control probability without increasing the healthy tissue complication probability new fractionation schemes have been proposed. The development of new fractionation schemes was strongly influenced by the so called linear-quadratic model.

4 Linear-Quadratic Model

In the previous section it was proposed to calculate the surviving fraction with the formulae:

$$SF(d) = \exp(-\alpha \cdot d - \beta \cdot d^2) \tag{8}$$

For fractionated radiotherapy the irradiation is repeated several times, usually once a day, until the total dose is achieved. It has already been pointed out that each successive fraction is equally effective. Therefore after L fractions the cell surviving fraction can be expressed as:

$$SF(d,L) = \left(\exp(-\alpha \cdot d - \beta \cdot d^2)\right)^L = \exp(-\alpha \cdot D - \beta \cdot D \cdot d) \tag{9}$$

where D is the total dose, i.e. D = L · d

It seems rational to identify the clinical effect *E* of radiotherapy, e.g. the local control of a tumour or the probability of injury of the surviving fraction:

$$E(d,L) = \exp(-\alpha \cdot D - \beta \cdot D \cdot d) \tag{10}$$

Let us consider two fractionation schemes, the first one with a dose per fraction d1 and the total dose D1 and the second one with a dose per fraction d2 and the total dose D2. The clinical effect for the same tissue (tumour is also treated as a tissue) is numerically represented by the following formulas:

$$E1 = \exp(-\alpha \cdot D1 - \beta \cdot D1 \cdot d1) \tag{11}$$

$$E2 = \exp(-\alpha \cdot D2 - \beta \cdot D2 \cdot d2) \tag{12}$$

These two schemes are iso-effective if E1 = E2.

$$E1 = E2 \rightarrow \exp(-\alpha \cdot D1 - \beta \cdot D1 \cdot d1) = \exp(-\alpha \cdot D2 - \beta \cdot D2 \cdot d2) \tag{13}$$

This equation may be rearranged into the following form:

$$\frac{D2}{D1} = \frac{d1 + \alpha / \beta}{d2 + \alpha / \beta} \qquad (14)$$

This equation was proposed by Withers in 1983 and is successfully used in clinical practice. One year earlier Thames published a paper presenting a survey of iso-effect curves for various normal tissues, mainly for mice. The most important conclusion from his work was that the iso-effective dose increases more rapidly with a decreasing dose per fraction for late effects than for acute effects. The distinction between early (acute) and late effects and early and late responding tissues is not simple. The most straightforward division on early and late effects is based on the time which passed between the start of radiotherapy and the manifestation of injury. The early reactions usually take place during the treatment or shortly after completing the treatment (within 90 days after the start of radiotherapy). Typical examples include mucositis and dermatitis. In most cases early reactions are transient. Late effects occur several months or even years after radiotherapy. Typical examples are myelopathy, fibrosis and necrosis. Unfortunately, in most cases the late injuries are irreversible and the severity of them may even increase with time.

To understand how Thames' observation might be described by the linear-quadratic model, let us compare the second fractionation scheme (fraction dose d and the total dose D) with a reference one. In radiotherapy practice the fractionation scheme with 2 Gy dose per fraction is commonly used. Therefore, it is very convenient to compare the new fractionation scheme to the 2 Gy one. Let us assume, that for the reference scheme the clinical effect is E. Then it is possible to rewrite formulae 13 in the form:

$$D = \frac{E}{1 + \frac{d}{\alpha / \beta}} \qquad (15)$$

The interpretation of the above formulae is very simple. The isoeffective total dose depends on the fraction dose d (this is obvious) and on the parameter α/β only (!). The dependence is more pronounced if the α/β parameter is small in comparison with the dose per fraction d. If the α/β parameter is much higher than the dose per fraction, then the total dose does not depend on the dose per fraction at all. The analysis of the results of treatments with different fractionation schemes reveals that for acutely responding tissues (early reactions) the α/β ration is within range 7–20 Gy, while for late responding tissues the α/β parameter is much smaller, in most cases smaller than 6 Gy. The malignant tissue is thought to be similar to early responding tissue with the α/β about 10 Gy, however there is some quite new data indicating that for some tumours the α/β parameter possibly is much smaller, within the range typical for the late responding tissues. One should emphasise that the uncertainties of α/β values are rather large. Therefore, the precision of iso-effective dose calculation is not very high. The second important remark is related to the fact that in the model formulated by Withers, time is not accounted for.

Therefore, the general rules, which should be followed whenever the linear-quadratic model is used for comparing two different treatment schedules, are: (1) the total treatment time in both treatment schedules should not be very different, (2) there should be a sufficient time between fractions for complete repair of sublethal damage, i.e. at least 7 h. The last remark we would like to make is that the range of fraction doses for which the model is valid has not yet been established. It is very likely that the model may be used safely for fraction doses within range of 1.4–5.0 Gy. In other words, the model is very attractive but it should be used with great caution. Let us apply the model for some clinical situations.

4.1 The Linear-Quadratic Model in Practice

Example 1

A patient has a metastasis in the thoracic vertebra. To make the treatment as convenient as possible for the patient a palliative, short radiotherapy is planned using 5×4 Gy. From clinical experience it is known that the safe dose delivered to the spinal cord is 45 Gy in 2 Gy per fraction. While using the linear quadratic model one should check whether this short fractionation is safe for the patient. Let us assume that the α/β parameter for the spinal cord myelopathy is 2 Gy.

To answer the question let us calculate the maximum safe dose delivered to the spinal cord in 4 Gy per fraction. From Eq. 13 we get:

$$D_{4Gy}^{safe} = D_{2Gy}^{safe} \cdot \frac{2Gy + \alpha / \beta}{4Gy + \alpha / \beta}$$

$$D_{4Gy}^{Safe} = 45Gy \cdot \frac{2Gy + 2Gy}{5Gy + 2Gy} \approx 25.7Gy$$

The maximum safe dose in 4 Gy per fraction is 25.7 Gy. This is larger than the planned total which is 20 Gy and therefore the patient may be treated with this fraction schedule.

Example 2

The conventional treatment technique for patients with head & neck cancers was based on the application of two parallel opposed beams. In the first phase of the treatment patients were irradiated with the so called large fields which enveloped both tumour and regional lymph nodes. Unfortunately, the high dose region usually encompasses the spinal cord, which absorbs the dose very similar to the fraction dose, so the full dose could not be delivered in this way due to the risk of myelopathy. After the spinal cord received the maximum safe dose, in the second phase of treatment, the field sizes were diminished to omit the spinal cord. If needed, the lymph nodes were irradiated with electron beams. As it was said in Example 1 the tolerance dose for the spinal cord is 45 Gy in 2 Gy fractionation regime. The 45 Gy could not bc dclivered in the fraction dose of 2 Gy, and therefore the 44 Gy were usually delivered in 22 fractions. Let us consider a following situation. The team which prepared the treatment plan made a mistake. Instead of 2 Gy fraction dose the patient was treated with fraction dose of 2.4 Gy. The mistake was discovered after ten fractions. What extra dose may be safely delivered to this patient with 2 Gy fraction schedule?

It is very easy to show that Eq. 13 may be written in the form:

$$\exp(-\alpha \cdot D1 - \beta \cdot D1 \cdot d1) = \exp(-\alpha \cdot D2 - \beta \cdot D2 \cdot d2) \cdot \exp(-\alpha \cdot D3 - \beta \cdot D3 \cdot d3) \quad (16)$$

In the Eq. 16 D2, d2 and D3,d3 denote the total as well as the fraction doses to the wrongly performed treatment and for the treatment after the correction of the error, respectively. We may compare two fractionations schemes composed of different doses per fraction. Equation 16 may be arranged into the following form:

$$D1 \cdot (d1 + \alpha / \beta) = D2 \cdot (d2 + \alpha / \beta) + D3 \cdot (d3 + \alpha / \beta)$$

As in Example 1, we choose for myelopathy $\alpha/\beta = 2$ Gy. Inserting the values of the total dose and dose per fraction for the planned treatment (44 Gy, 2 Gy), and for the treatment in which the mistake was made (24 Gy, 2.4 Gy) we have:

$$44Gy \cdot (2Gy + 2Gy) = 24Gy \cdot (2.4Gy + 2Gy) + D3 \cdot (2Gy + 2Gy)$$

The solution of this equation gives the value 17.6 Gy for D3. It is very likely that the extra nine fractions with 2 Gy dose per fraction (total dose of 18 Gy) would be delivered before the beam size would be reduced.

5 Summary

Radiobiology, as a separate branch of biological science, played an important role in the development of rational radiotherapy. Its role seems to be even more important today in the era of application of protons and heavy ions in radiotherapy. More successful radiotherapy, and in general more successful treatment of patients with cancer, comes through better understanding of the interaction of ionizing radiation with tissues. The involvement of physicists in radiotherapy demands at least some understanding of complicated relationships between the delivered radiation dose and treatment outcome. History showed that the physicist's skills in mathematical modelling of different processes lead to many practical applications of radiobiology in radiation therapy.

References

1. E. Podgorsak, Radiation Oncology Physics: A Handbook for Teachers and Students, IAEA, Vienna. www.pub.**iaea**.org/MTCD/publications/PDF/Pub1196_web.pdf.
2. A. Van der Kogel and Michael Jioner, Basic Clinical Radiobiology, A Hodder Arnold Publication, London, 2009.

Part III
Radiation Protection Application in Hospital Medical Physics

Part III
Radiation Protection Applications
in Diagnostic Medical Physics

General Principles of Risk Assessment for Ionising Radiation

Jim Malone and Geraldine O'Reilly

Abstract This paper provides a general background to the topic of Risk Assessment with a view to applying it helpfully in ensuring safe application of Ionizing Radiations in Diagnostic Medical Imaging. Similar principles can be applied in Radiation Therapy.

Keywords Risks • Risk assessment • Radiation • Equipment

1 Background to Risk Assessment

This lecture provides a general background to the topic of Risk Assessment with a view to applying it helpfully in ensuring safe application of Ionizing Radiations in Diagnostic Medical Imaging. Similar principles can be applied in Radiation Therapy.

A formal approach to risk assessment has been found useful in the management of many types of hazard, and helps anticipate serious risks before they manifest themselves through injury of individual workers, patients or members of the public. The extent to which the risk assessment approach is built into the regulatory and licensing requirements of individual European countries varies considerably. Nevertheless the key elements of the approach can be successfully applied in widely differing frameworks. This paper describes how it is applied in one country, and emphasises points of general applicability.

In Ireland the requirement for a Risk Assessment arises as part of the licensing procedure for equipment, facilities and use of particular radionuclides. In a Risk Assessment, the employer seeks to identify ***hazards,*** the likelihood of injury arising from them, and put control measures in place to limit their impact. Hazards are loosely defined as ***situations in which there is a potential for human injury or***

J. Malone (✉) and G. O'Reilly
Department of Medical Physics and Bioengineering, Trinity College/St James's Hospital, Dublin 8, Ireland
e-mail: jifmal@gmail.com; jfmalone@tcd.ie

Y. Lemoigne and A. Caner (eds.), *Radiation Protection in Medical Physics*,
NATO Science for Peace and Security Series B: Physics and Biophysics 1,
DOI 10.1007/978-94-007-0247-9_5,

Table 1 Prevention and reduction of risk

Nine principals of prevention
Avoidance of the risk
Evaluation of unavoidable risk
Combating the risk at source
Adapting the work to the individual
The adaptation of the workplace to technical progress
The replacement of dangerous substances or systems of work
Give priority to collective protective measures
The development of prevention policies
Training and PPE

damage to the environment or both. Clearly ionizing radiation is a potential hazard within this definition, although it shares this with many other hazards such as chemical, infection and fire hazards in hospitals and clinics. Risk Assessments, through identifying hazards seek to establish control measures which will, where possible eliminate them and where they cannot be eliminated, they will be reduced to an acceptable level and their impact minimized. There are many approaches to this, including use of the nine principles of prevention in Table 1.

Medical Physicists play a key role in practice in creating risk assessment frameworks for ionizing radiation in diagnostic imaging facilities, providing templates against which potential risks can be assessed, and implementing solutions to control the hazards identified. Physicists tend to be excellent at identifying technical problems, such as whether or not a wall is adequately shielded, or if equipment is working properly. However, their input is also a need to prior questions, such as if the framework for management of radiological safety issues is broad enough and robust enough to withstand serious problems, and if the staff involved are well enough trained and motivated to provide the operational support necessary. These and some other basic concerns are just as important as the technical issues involved.

2 Examples of Elements from Risk Assessments

The associated lecture deals with examples of risk assessments in diagnostic radiology and nuclear medicine by looking at both generic and specific problems. Table 2 on the previous page illustrates three of 25 areas examined when a risk assessment is being conducted for an installation in a diagnostic radiology room.

The assessment is performed for a department with an exceptionally heavy workload in general radiology and CT. The example is a real one in a hospital with an exceptionally high workload for the facilities it enjoys, but an otherwise good practice in radiation protection.

Table 3 provides an illustration of ten specific examples of external irradiation hazards that arise in diagnostic nuclear medicine. The risk analysis must provide an

Table 2 Some questions raised during a risk assessment

System		General (A&E)	CT
Question 1:	Are the walls well protected?		
Risk	Exposure to persons on other side of wall	Response: Measurements show at least 2 mm lead equivalent shielding in all walls with the exception of the wall to the toilet, which is slightly less attenuating	Response: Internal & external walls min. 215 mm thick solid concrete. 25 mm Barium plasterwork on walls.
People at risk	Designated and non-designated staff, patients, members of public	Follow-up: None based on current workload, however the shielding of the toilet wall must be reviewed if the occupancy or usage changes. A beam blocker is advised behind the erect Bucky (see below)	Follow-up: None
Control:	Design and commissioning of X-ray department		
Responsible person:	RPA		
Question 4:	Is door lead-lined?		
Risk	Exposure to persons outside door	Response: Attenuation measurements indicate that doors are shielded but less than 2 mm lead equivalent. No shielding in doorframes.	Response: All doors, frames and architraves lined with 2.24 mm lead.
People at risk	Designated and non-designated staff, patients, members of public	Follow-up: Doors & doorframes to be upgraded by addition of 1 mm lead or replaced with doors & frames of 2 mm lead equivalence	Follow-up: None
Control:	Design and commissioning of X-ray department		
Responsible person:	RPA, Hospital radiation safety committee		
Question 8:	Do only suitably qualified people operate equipment?		
Risk	Risk of over-exposure	Response: Yes	Response: Yes
People at risk	Designated staff, patients	Follow-up: None	Follow-up: None
Control:	Check staff qualifications; Limit access to qualified people only		
Responsible person:	Radiographic service manager; Radiologist in charge; Hospital radiation safety committee		

adequate means of dealing with each of these, and of dealing with the other hazards involved, including contamination.

In the associated lecture, further examples are provided drawing on assessment of shielding, buildings, building materials employed, windows, installation features, space and layout, control of access, operator performance, protocols, and protective

Table 3 External radiation hazards in nuclear medicine

1.	External irradiation: sealed radioactive sources used for testing
2.	External Irradiation: technetium generators whilst in use or storage
3.	External irradiation: unsealed radionuclides being manipulated or in storage
4.	External irradiation: sources/radionuclides being transported to site of usage
5.	External irradiation at site of use
6.	External irradiation: patients to whom radiopharmaceuticals administered
7.	External irradiation from spent radioactive sources
8.	External irradiation from used syringes, vials, swabs or other waste
9.	External radiation from tissue samples during surgical procedures
10.	External irradiation from blood samples

devices/accessories. In addition some features of the main types of specialized equipment employed are examined, including general radiography, dental units, mammography, fluoroscopy, specialised interventional equipment, CT scanners, and nuclear imaging devices.

References

1. BIR (2000). Radiation Shielding for Diagnostic X-rays, Report of a joint BIR/IPEM working party. Edited by D.G. Sutton and J.R. Williams. London: British Institute of Radiology.
2. RPII (2004). Guidance Notes on Radiation Risk Assessment. Dublin: Radiological Protection Institute of Ireland.
3. Dublin Hospitals Risk Management Forum (2002). Report of Sub-Committee dealing with Radiation Safety, Final Report, July 2002.
4. Pearson D, Rogers AT and Moss E (2001). A generic approach to risk assessment for the Ionising Radiations Regulations 1999. Br J Radiol, 74, 62–68.

Radiation Protection of the Workers and the Public from Exposed Patients

Jim Malone

Abstract Radiation protection is concerned with reducing the dose to patients, workers, and other third parties such as comforters/carers and members of the public. The latter include workers in hospitals and clinics that are not recognised as radiation workers. This paper focuses on protection of workers and other third parties from radiation that arises from conduct of diagnostic imaging procedures.

Keywords Radiation protection • Workers • Public • Patients • Scattered radiation • Radiology • Interventional • Cardiac

1 Introduction

Radiation protection is concerned with reducing the dose to patients, workers, and other third parties such as comforters/carers and members of the public. The latter include workers in hospitals and clinics that are not recognised as radiation workers. This paper focuses on protection of workers and other third parties from radiation that arises from conduct of diagnostic imaging procedures. In practice such radiation frequently originates from the patient. This is so in the case of diagnostic radiology as third parties will seldom, with good practice, find themselves in the direct beam. Instead most of their exposure will arise from secondary radiations principally that scattered from the patient and to a lesser extent the leakage from the x-ray tube housing. In nuclear medicine imaging, the patient is obviously the main source of radiation for those he/she comes into contact with. Doses arising from radiopharmacy are not considered here.

J. Malone (✉)
Department of Medical Physics and Bioengineering, Trinity College/St James's Hospital, Dublin 8, Ireland
e-mail: jifmal@gmail.com; jfmalone@tcd.ie

Y. Lemoigne and A. Caner (eds.), *Radiation Protection in Medical Physics*,
NATO Science for Peace and Security Series B: Physics and Biophysics 1,
DOI 10.1007/978-94-007-0247-9_6,

2 Practical Protection in Radiology

In diagnostic radiology, much of the protection of third parties, including workers and the general public arises from well designed tube housing, collimation arrangements for x-ray tubes, and structural shielding. The radiation not contained by the tube housing and collimator generally impinges on the patient and a significant proportion will be scattered. The room structural shielding will contain the scattered radiation, thereby protecting all those outside the room. However those inside the room need added protection. In addition some practices need special additional precautions. These include situations where a mobile unit or mobile C-arm is used outside a room with structural shielding. Special considerations may also apply to equipment temporarily provided on a trailer parked in the hospital or clinic car park.

Medical Physicists have an important role in facilitating safe working practices for radiation workers such as radiologists, cardiologists, radiographers, technologists, nurses and others. Their work may be undertaken in the context of a regulatory environment or of a hospital department. The concern of a regulatory environment may be compliance. However, for physicists working in a hospital there should be a much greater emphasis on facilitating the work of the institution.

Facilitating the work of the groups listed requires an understanding of their needs. It is important to be aware that they generally have little real interest in the dose registered in personnel monitoring systems, as the units and quantities employed to communicate dose, by and large, lack transparency outside the physics world. Topics they may be interested in, are illustrated in Table 1. To be genuinely helpful it is necessary to be aware that most of these workers operate in a multi-hazard environment which includes serious infection control and pharmaceutical toxicity concerns. Thus the special issues associated with radiation are one of many concerns they will have. It is also necessary to be aware that some radiation protection advice can have adverse effects. For example, Pb aprons can give rise to low back pain and related cervical and lumbar spine problems. Likewise, some workers find Pb glasses to be heavy, uncomfortable and to restrict peripheral vision. Many of the measures proposed to reduce radiation are uncomfortable for those involved, distract them, impede free movement and reduce performance. It is necessary to be

Table 1 Radiation workers want to know[a]

Radiation workers want to know
Radiation dose, little interest
What does the dose number mean?
What is my risk?
What is the risk to my baby?
Is my work practice ok?
Is there anything I should change?

[a]This table and some of the related text are based on observations made by Dr. D. Millar during a lecture at the IAEA in Vienna, and is confirmed by the author's experience

aware of these limitations, something which is readily achieved by wearing the protective systems for at least a full working day.

One of the more common situations where the operator and support staff must be in the room with the patient while the beam is on is during fluoroscopic and interventional procedures. In such situations there is a direct relationship between patient and staff doses, thereby providing the operators with a strong incentive to reduce both. Positioning the operator and the equipment correctly and use of appropriate exposure factors has an decisive influence on both patient and staff doses.

Practical suggestions that greatly assist radiation protection are listed in Table 2. Possibly the most important choice made by the operator is the "mode" or protocol used. The options include low and high dose fluoroscopy, pulsed mode operation and "fluorography" or "imaging" (which may be similar to the older cine fluorography in cardiology). The difference in dose rate per image between the highest and lowest modes is several orders of magnitude, and the modes are sometimes used without an awareness of this. Terminology is not standardised between suppliers, and thus there is much scope for very large exposures being employed when they are not needed.

It is particularly important to have the image receptor as close as possible to the patient and to keep the distance between the image intensifier and the patient as large as is allowed by the equipment design and the ergonomic situation. When an undercouch tube configuration is employed the operator will receive large doses to the lower limbs if protective clothing and/or table side drapes/enclosures are not provided. Use of an over-couch tube gives large upper body, head and neck doses to the operator(s). It is essential to protect against these. The measures required involve a mixture of minimising the beam on time and current, standing back, remote injection systems, ceiling suspended table side shielding, and personal protection systems including aprons, glasses, etc. (Fig. 1)

The layout and design of a room, as well as structural shielding do much to protect workers and the public (structural shielding) and the patient, comforters/carers and radiation workers (layout, design and equipping of the room). Special features help optimise design for each imaging modality including radiography, dental, mammography, use of mobile devices, fluoroscopy, special/interventional procedures, and CT. These issues are also briefly addressed in the chapter on shielding. Protective devices and clothing add to the structural and layout features of an installation, and

Table 2 Summary guidelines for fluoroscopy/interventional

Good practice with fluoroscopy
Exposure protocol and beam factors/settings are crucial
Use pulsed mode where practical
Multidisciplinary approach essential
Keep the II close to the patient
Do not overuse magnification modes
Keep the x-ray tube at maximal distance from patient
Use higher kVp where possible
Wear protective aprons and radiation monitors
Know where scatter is highest
Keep your distance, as far as is practicable

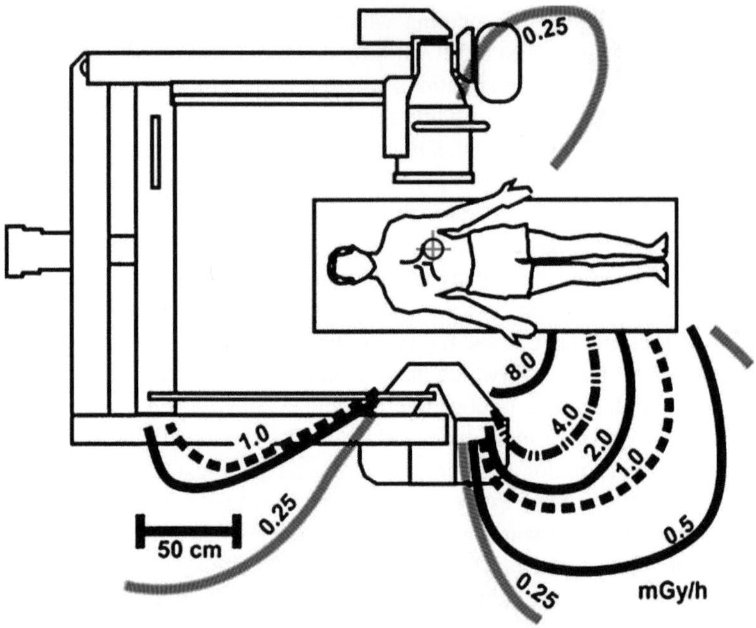

Fig. 1 Dose rates 100 cm above ground with image intensifier and x-ray tube in horizontal plane (Courtesy of Dr Steve Balter)

include, the operator's console area, ceiling suspended protective lead glass screens, mobile screens, lead curtains and blinds, lead aprons, spectacles and goggles, a thyroid collars. The effectiveness of all these measures should be confirmed by monitoring. Additional useful information available is available in the lectures on the IAEA web site at:http://rpop.iaea.org/RPOP/RPoP/Content/AdditionalResources/Training/1_TrainingMaterial/index.htmand from other sources.

Installation Shielding in Diagnostic Radiology and Computed Tomography

Jim Malone, Una O'Connor, and Noirin Sheahan

Abstract This paper deals with shielding of diagnostic radiological facilities to produce an environment that is safe for workers, patients and the general public. The radiations involved will range from dental and DXA at one end to CT and interventional radiology at the other. The problem of identifying the goal of shielding design, which varies in different parts of the world, will be presented and solutions suited to Europe will be identified. Examples of designs based on these goals will be presented.

Keywords Radiology • Shielding • Dose reduction • Staff safety • Dose constraints

1 Introduction

This paper deals with shielding of diagnostic radiological facilities to produce an environment that is safe for radiation workers, other workers, patients and the general public. The radiations involved range from dental and DXA at one end to CT and interventional radiology at the other. The problem of identifying the goal of shielding design, which is different parts of the world, is presented and solutions suited to Europe are identified. Examples of designs based on these goals are presented. The key sources of information available in the area are identified and briefly reviewed.

J. Malone (✉), U. O'Connor, and N. Sheahan
Department of Medical Physics and Bioengineering, Trinity College/St James's Hospital, Dublin 8, Ireland
e-mail: jifmal@gmail.com; jfmalone@tcd.ie

Y. Lemoigne and A. Caner (eds.), *Radiation Protection in Medical Physics*, NATO Science for Peace and Security Series B: Physics and Biophysics 1,
DOI 10.1007/978-94-007-0247-9_7, © Springer Science+Business Media B.V. 2011

2 Some Practical Considerations

The types of installation encountered in diagnostic radiology include those in Table 1. The unit type is highly diverse and the designs for safe shielding solutions are equally diverse. Even greater diversity is encountered in nuclear medicine and radiotherapy.

The radiation types for which shielding must be designed are classified into two broad categories, primary and secondary radiations. The primary beam is the most energetic form of radiation encountered in diagnostic radiology and is the beam as it emerges from the tube head. In many diagnostic applications, this is intercepted by the patient, the x-ray table, the image receptor and beam blockers and will not present a problem to be dealt with by structural shielding. However, there are occasions where the direct beam must be intercepted and attenuated by structural shielding and examples are presented in the references. Secondary radiations consist of radiation scattered (most from the patient), and other radiations such as tube leakage. With well designed modern equipment the former tends to dominate except at very short distances from the tube housing. The implications of both radiation types for the design and layout of radiological facilities and for shielding calculations are important.

Good layout of appropriate sized rooms greatly improves staff safety and reduces the need for structural shielding. Examples of good room layout are given in Fig. 1 for a general radiography room and for a CT room in Fig. 2. Examples of other room designs are given in RPII [4].

Calculations are generally performed using methodologies published by the NCRP in the US, or by the BIR in the UK. A recent review of both methods is available with commentary well suited to their current application in Europe. The fundamental physics of both methods is the same, but in practice they can yield different solutions. The principal differences in the methods (Table 2) relates to the dose constraint used to identify the objective of shielding calculations, and the assumptions employed in respect of the workload in the facility and occupancy/ use of adjacent areas. The RPII [4] approach recommends a dose constraint of 0.3 mSv per year, which is emerging as a value used by several states in Europe. With respect to workload and other factors, it recommends use of real data, where

Table 1 Types of facility in diagnostic radiology

Types of facility
Dental units
General radiology
Mammography units
Chest units
General fluoroscopy units
Interventional units
Mobile general units
Mobile C-arms
CT units
Units on trailers

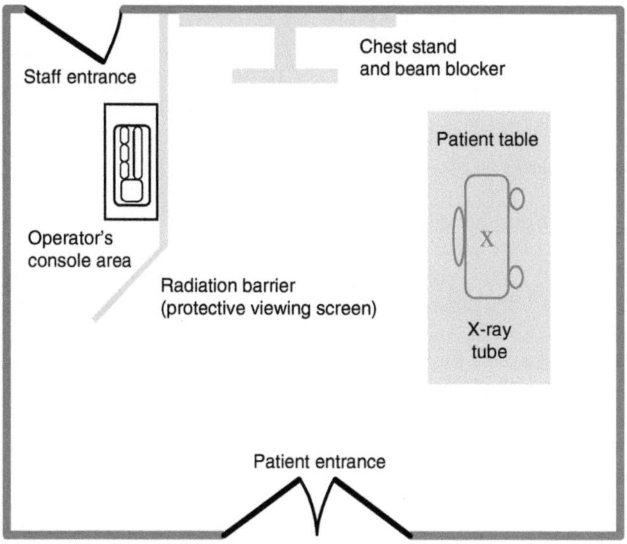

Fig. 1 A good general radiology room design (RPII [4])

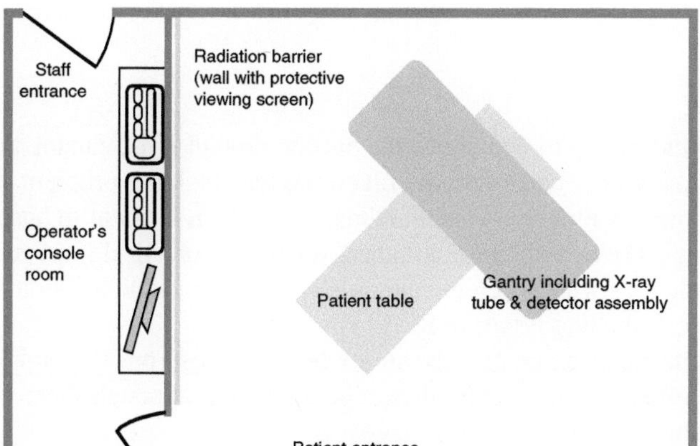

Fig. 2 Example of good CT room design (RPII [4])

available, as opposed to the values tabulated in the NCRP and BIR approaches. Both are primarily for use in their countries of origin, are untested in other countries and, in important cases, may not represent current practice.

Traditionally the shielding required in diagnostic radiology was fairly predictable and there were few situations that could not be adequately dealt with by 2 mm of Pb or equivalent. However this has changed and significantly more shielding is normally required for modern multislice CT installations with a full workload. In addition rules of thumb that were previously employed, such as shielding external

Table 2 Comparison between BIR and NCRP shielding methodologies. Based on BIR [1] and NCRP [2], and RPII [4]

Shielding concept	BIR methodology	NCRP methodology	Comments
Classification of areas	Controlled	Controlled	Controlled
	Not specified	Uncontrolled	Public area and all others[a]
Design limits terminology	Dose constraint	Shielding design goal	Dose constraint
Design limits employed	Not specified	5 mSv/year (Controlled)	RPII, 1 mSv/year (Exposed workers)
	0.3 mSv/year (Public and non-radiation staff)	1 mSv/year (Uncontrolled)	RPII, 0.3 mSv/year (All others)[a]
Weekly workload (Primary)	Entrance Surface Dose (ESD) or Film dose	mA min	Use real data if possible, otherwise either or both
Weekly workload (Secondary)	Dose-Area Product (DAP)	mA min	Use real data if possible, otherwise either or both
Occupancy	Percentage of time	Fraction of time	Use real data if possible, otherwise either or both

[a] A conservative approach requires use of a dose constraint of 0.3 mSv/year for supervised areas, as there can be exposed workers, non-exposed workers and members of the public present in these areas

walls and windows up to 2 m above the outside ground level, cannot any longer be safely employed on a routine basis. Likewise, the shielding problems that prevail when equipment is mounted in trailers that travel from hospital to hospital require new solutions. These problems and others relating to practical implementations of shielding solutions with different materials are well treated in a way that addresses contemporary shielding issues in RPII [4].

Finally, the balance that must be struck between, cost, practical implementation issues, and effectiveness should always be addressed although there is little published literature on this.

References

1. British Institute of Radiology (BIR). (2000). Radiation Shielding for Diagnostic X-rays, Report of a joint BIR/IPEM working party. Edited by D.G. Sutton and J.R. Williams. London: BIR.
2. National Council for radiation Protection (NCRP). (2004). Structural Shielding Design for Medical X-Ray Imaging Facilities, Report no. 147. Bethesda: NCRP.
3. National Health Service (NHS). (2001). Facilities for Diagnostic Imaging and Interventional Radiology, HBN 6, NHS Estates, London Stationery Office.
4. Radiological Protection Institute of Ireland (RPII). (2009). The Design of diagnostic medical facilities where radiation is used. Dublin: RPII. www.rpii.ie.

Acceptability Criteria, Justification and Ethical Issues in Using Ionising Radiation

Jim Malone, O. Holmberg, and R. Czarwinski

Abstract Radiation Protection in medicine is underpinned by the concepts of **Justification** and **Optimization**. Both IAEA and EC initiatives now emphasise the need to address the area of Justification urgently. Ethical performance of examinations also requires that they be undertaken on equipment suited for that purpose.

Keywords Radiation • Justification • Referral guidelines • Clinical audit • Criteria for acceptability of equipment • Ethical issues in radiology

1 Justification and Ethical Issues

Radiation Protection in medicine is underpinned by the concepts of ***Justification*** and ***Optimization***. Over the last 20 years much successful work has been devoted to developing and consolidating approaches to optimization. Less effort has been committed to justification and those applied have not yet been as successful. Both IAEA and EC initiatives emphasise the need to address the area urgently. Authoritative sources suggest that a substantial fraction (20–40% routinely in good departments, with up to 75% in some areas) of radiological examinations may be unnecessary. Current experience and the published literature suggest that in many clinical settings, both referring physician and imaging practitioners have limited awareness of the actual doses and risks involved. For example a mainstream general medical journal,

J. Malone (✉)
IAEA, Vienna, Austria
and
Department of Medical Physics and Bioengineering, Trinity College/St James's Hospital, Dublin, Ireland
e-mail: jifmal@gmail.com; jfmalone@tcd.ie

O. Holmberg and R. Czarwinski
IAEA, Vienna, Austria

Y. Lemoigne and A. Caner (eds.), *Radiation Protection in Medical Physics*,
NATO Science for Peace and Security Series B: Physics and Biophysics 1,
DOI 10.1007/978-94-007-0247-9_8,

identified that few of those responsible for prescribing or performing examinations are familiar with the units used to specify the amount of radiation received, and the risk associated with it. These observations have been confirmed at consultations held by the IAEA and a joint Workshop held by the IAEA and the EC in 2009.

These findings raise serious ethical, legal and practical concerns. The ethical and legal concerns derive from the view of the person out of which radiation protection is working. It is felt the model is, perhaps, too paternalistic and does not take sufficient account of the autonomy and dignity of the individual. Thus there is a greater need to be aware of the patient and his or her wishes in the justification process. Informed consent gains a new importance, and examinations should be undertaken in a way that is transparent and accountable to the patient as well as to the professions. This will place additional burdens on those involved in the practice of radiology and radiation protection.

Consultations on practical measures that can improve the effectiveness of implementation of justification in the day to day practice of hospitals and clinics identified a number of key practical issues. They are a means of ***ensuring that those referred for radiological examinations really need them***; the ***audit of the effectiveness of the referral and related processes***; and finally devising means of ***effectively communicating about radiation risk*** to patients, physicians, surgeons, allied professionals, and of course the radiologists who are generally responsible for performing them. Tools for each of these tasks have been explored in the literature. They include referral guidelines, clinical audit and recognition of the failure of the system of radiation units as a means of communicating about risk to medical and other health professionals.

Referral Guidelines for radiological and nuclear medicine examinations have been issued by the Royal College of Radiology in London (RCR), the American College for Radiology (ACR), the EU and others. In Europe, these are required by the MED (Medical Exposures Directive). The chief causes of unnecessary or wasteful radiology identified by the RCR are:

- Repeating Investigations already done.
- Undertaking investigations unlikely to influence patient management.
- Investigating too early.
- Doing the wrong investigation.
- Failing to provide clinical information and questions to be answered.
- Over investigation.
- Other aspects of potentially unsatisfactory referral patterns have been noted in the report of an IAEA consultation held in 2007, and include non medical referrals, self referral, self presentation, some screening programmes, and referrals arising from social, economic or political pressures.

Within modern medicine, audit is a key component of all disciplines, and is now being introduced in diagnostic radiology. The EU MED Directive requires that clinical audits be performed and the development of useful audit tools is underway. The evidence available indicates that clinical audit is a simple and effective method for improving referral patterns.

The IAEA consultations recognised the failure of successive approaches to communication of radiation risks to various groups including patients, practicing physicians and surgeons, allied professionals, and radiologists. The importance of a more effective approach in this regard has been given additional weight by recent communications and debate in both the specialist medical literature and in the public press. While communication about risk is central to the above problems, the manner in which communication with patients and between professionals is undertaken is central to these questions also. This will not be further treated here other than to note the importance of the issue and the fact that it has been acknowledged.

2 Criteria of Acceptability for Equipment

In addition it has been recognised that ethical performance of examinations requires that they be performed on equipment that is acceptable from the point of view of patient safety and adequate performance of the examination. The EC has commissioned a revision of the Criteria for the Acceptability of Equipment in Radiology, Nuclear Medicine and Radiotherapy. The current draft of these, which is expected to be available for public consultation at the time of the 2009 school, will also be reviewed.

The EU Medical Exposures Directives (MED) require the establishment of criteria of acceptability of radiological, nuclear medicine and radiotherapy installations. To assist the member states the European Commission published RP91, "Criteria for Acceptability of Radiological (including Radiotherapy) and Nuclear Medicine Installations" in 1997. In 2007, a group was commissioned to revise RP91; the final draft should be available for public comment at the time of the 2009 school.

A critical reading of the MED, RP 91, and the professional literature reveals some shift or "creep" in the meaning of the terms remedial and suspension level since they came into widespread use in the mid 1990s. In the interest of clarity, the Suspension Level has been redefined as follows:

A level of performance that requires the immediate removal of the equipment from use.

The new draft also indicates that: *Following a documented risk assessment involving the MPE and the practitioner, the suspended equipment may be considered for use in limited circumstances. The holder and operators must be advised in writing of the suspension and/or limitations in use.*

Thus failure to pass a suspension level requires that the equipment be taken out of service immediately. Such equipment would be unsafe, or its performance would be so poor, that it would be unacceptable to society. The level is based on meeting, at least, the minimum standards of safety and performance that would be acceptable in the EU. The level chosen represents the expert judgement of the working group and reviewers based on the knowledge of what is acceptable amongst their peers. It is also informed by social, legal and political circumstances that prevail in the EU.

It was not possible to devise a single acceptable approach to proposing values or levels for the criteria selected. Instead a number of approaches, with varying degrees of authority and consensus attached to them, have been adopted and grouped under headings A to D as follows:

Type A Criterion: is based on a formal national/international regulation or an international standard.
Type B Criterion: is based on formal recommendations of scientific, medical or professional bodies.
Type C Criterion: is based on material published in well established scientific, medical or professional journals.
Type D Criterion: arises where it has not been possible to make a recommendation.

The small residue of areas in which the Type D approach is required includes, for example, fields where the technology involved is evolving rapidly. Here, providing a value could be counterproductive. It could become out of date very rapidly or it could act as an inhibitor of development. In these and other Type D situations the criterion of acceptability should be determined by the institution holding the equipment based on the advice of the Medical Physics Expert or Radiation Protection Adviser as appropriate. Additional advice is provided in the revised draft on how the criteria should be applied in practice and how acceptable equipment should be signed off.

References

1. IAEA, Report of a consultation on justification of patient exposures in medical imaging, Radiat Prot Dosimetry 2009 135: 137–144; doi:10.1093/rpd/ncp107 Accessed 10th Aug 2009.
2. European Commission (2000). Referral Guidelines for Imaging. 2000. http://ec.europa.eu/energy/nuclear/radioprotection/publication/doc/118_en.pdf (Accessed 10 Aug 2009).
3. European Commission (1997) Criteria for Acceptability of radiological (including radiotherapy and nuclear medicine installations RP 91. European Commission, Luxembourg).
4. J. F. Malone Radiation protection in medicine: ethical framework revisited Radiation Protection Dosimetry Advance Access published on March 5, 2009 Radiat Prot Dosimetry 2009 135: 71–78; doi:10.1093/rpd/ncp010 Accessed 10th Aug 2009.

Dosimetry of Ionizing Radiation: In Search of an Ideal Detector

Pawel Kukolowicz

1 Introduction

Proper dose measurement skills are of the utmost importance for all applications of ionizing radiation in medicine. For years, since the discovery of ionizing radiation, the delivered dose to exposed people has been evaluated by means of subjective methods. In radiotherapy, the unit "erythema dose" was widely used. The erythema dose was connected to the reaction of the skin to radiation. (Strictly: The erythyma dose is the amount of radiation which, applied to the skin, makes it turn temporarily red [erythematous]. Webster's New World™ Medical Dictionary, 3rd Edition) In the early days of the discovery of X-rays, the Roentgen radiation was commonly used. Luckily, the Roentgen radiation deposits the maximum energy to the surface, i.e. in the case of radiotherapy to the skin. Careful observation of the redness of the skin allowed therapists to finish treatment in the right time, before a serious injury of deeper anatomical structures would occur. However, this type of measurement of delivered dose was very imprecise. It depended on the individual reaction of each single person. In 1924 an objective method, namely the unit of radiation exposure, was introduced. This unit, the Roentgen, was internationally accepted in 1928 during the II International Congress of Radiology held in Stockholm. It was based on the measurement of the ionization of air exposed to ionizing radiation. At that time the method of the measurement of charge has been well developed. The method of dose measurement, based on charge measurement has been developing for years and today is considered as the most precise and the simplest method of dose measurement. Nevertheless, new detectors and new methods of dose measurement are still developing, and, what has to be emphasized, is that there is no ideal radiation detector for all applications of ionizing radiation. In this lecture the theory of dosimetry will be presented alongside the most often used detectors in medical practice.

P. Kukolowicz (✉)
Medical Physics Department, Holycross Cancer Center, Kielce, Poland
e-mail: pk@message.pl

Y. Lemoigne and A. Caner (eds.), *Radiation Protection in Medical Physics*, NATO Science for Peace and Security Series B: Physics and Biophysics 1,
DOI 10.1007/978-94-007-0247-9_9,

2 Important Quantities Used in Dosimetry of Ionizing Radiation

The events that result in radiation energy absorption, and consequently in biological effects, are immensely complex. To describe the deposition of radiation energy to a tissue we will limit our considerations to photons. The way in which the energy is deposited is illustrated in a simplified way in Fig. 1.

In the first step the photon usually collides with an electron, resulting in scattering photons and setting an electron into motion. While travelling through the tissue, the electron interacts with mater. During these interactions mostly ionizations and excitations of atoms take place. The scattering photon interacts with the matter in a similar way as the primary photon. The electron travels through matter until it loses all its energy. In understanding the process of radiant energy deposition it is crucial to notice that the transfer of energy from the ionizing particle does not have to lead to absorption in the place where interaction took place. Therefore, it is very convenient to distinguish between energy transfer and energy absorption. Usually, the large amount of energy is transferred from photons to high energy electrons and the latter loose kinetic energy in many single, low-energy interactions. This phenomenon makes dosimetry difficult because, although we may easily measure the transferred energy, it is more problematic to convert the transferred energy to energy absorption. The second problem is that in any interaction it is impossible to state exactly what energy will be transferred. However, after many interactions, one can calculate the average energy transferred and with application of an appropriate theory the average energy absorbed can be established. Let us assume that we know the number of photons (number of electrons) that impinge on a layer of matter.

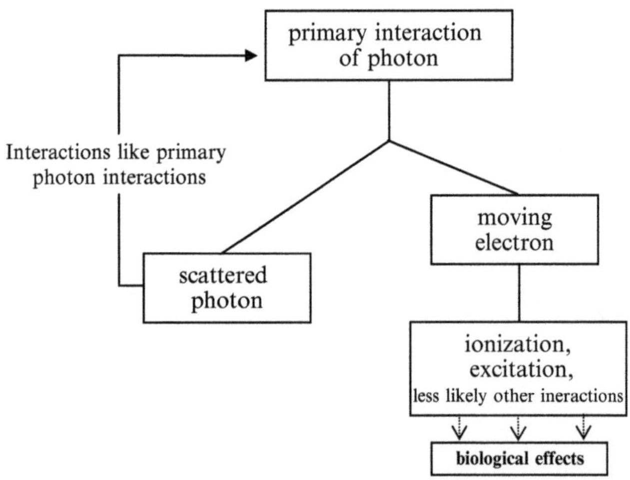

Fig. 1 Energy deposition

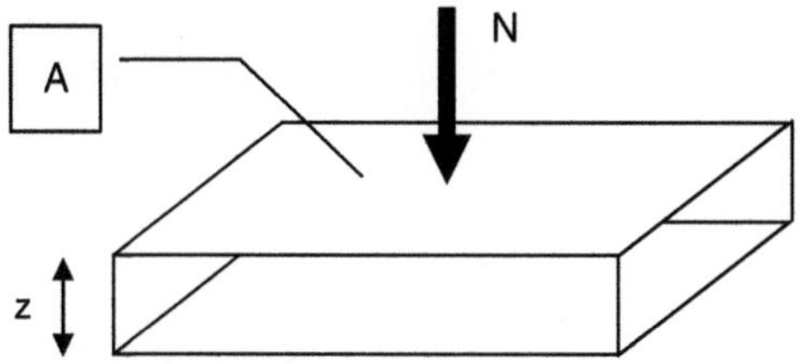

Fig. 2 Photon beam on a thin layer of material

Assuming that we know the number of photons impinging on a thin layer of matter let us calculate the energy transferred to a box of matter.

2.1 *Energy Transferred*

Figure 2 presents the photon beam impinging on a thin layer of, say, carbon. The area of this piece of absorber is A and the thickness is z. The number of photons impinging on the layer per 1 s is N. How many photons will interact with matter?

The number of interacting photons is proportional to N and to z. If N is doubled, then the number of interactions will also be doubled. If the thickness z is doubled so is the number of interactions. Thus, the number of interactions can be given by the formulae:

$$n = \mu \cdot N \cdot z \tag{1}$$

where μ is the constant of proportionality. This constant is called the linear attenuation coefficient.

If the average energy transferred per interaction is $E_{tr,avr}$ then the energy transferred is:

$$E_{tr} = \mu \cdot N \cdot z \cdot E_{tr,avr} \tag{2}$$

How much energy is transferred per unit mass? This is a very simple question. To calculate it we must divide the energy transferred to the block of Carbonits mass which is given by:

$$m = A \cdot z \cdot \rho \tag{3}$$

where ρ is the density of Carbon. The energy transferred per unit mass is given by:

$$\frac{E_{tr}}{m} = \frac{\mu \cdot N \cdot z \cdot E_{tr,avr}}{A \cdot z \cdot \rho} = \frac{\mu}{\rho} \cdot \frac{N}{A} \cdot E_{tr,avr} \tag{4}$$

The reader should notice that in our example the ratio N/A is the photon fluence (usually denoted with the capital Greek letter Φ). Therefore we may write the equation in the form:

$$\frac{E_{tr}}{m} = \frac{\mu}{\rho} \cdot \Phi \cdot E_{tr,avr} \tag{5}$$

The ratio of the linear attenuation coefficient and the density of an absorber is called the mass attenuation coefficient. This coefficient does not depend on the density of a material – one may say that the mass attenuation coefficient characterizes the material independently on external conditions. This is essential if the radiation interactions are measured in gases which change their density with pressure and temperature. The average energy transferred to electrons per unit mass is called Kerma (**K**inetic **E**nergy **R**eleased per Unit **Ma**ss).

2.2 *Energy Absorption – Absorbed Dose*

Part of the photon energy is transferred to electrons. However not all the kinetic energy of the electrons will be absorbed. Some energy will be irradiated in the form of Bremmstrahlung radiation. For photon beams of energy typical for medical applications less than a few percent is irradiated in the form of Bremmstrahlung.

Let us imagine a small volume dV of mass dm of any absorber and any type of ionizing radiation passing through it. The energy imparted **ε** to the volume dV is the difference between the sum of all energy entering the volume and the sum of all energy leaving the volume, taking into account any energy which is converted to mass. If in the volume an electron-positron pair is created, then the energy imparted should be decreased by the energy of mass of these two particles and vice versa. If electron-positron annihilation takes place in the volume dV, the energy imparted should be increased by the energy of the mass the said pair. The absorbed dose is the ratio of the mean energy imparted to this small amount of matter and the mass of this matter.

$$D = \frac{d\bar{\varepsilon}}{\rho \cdot dV} = \frac{d\bar{\varepsilon}}{dm} \tag{6}$$

The unit of absorbed dose is the joule per kilogram (J/kg). The name for the unit of absorbed dose is the gray (Gy). The term "absorbed dose" is used for describing the energy deposited in a point. According to definition "the point" should be large enough to consider the interactions as non-stochastic process.

2.3 *Stopping Power*

We have already gained some knowledge on how photons deliver the radiant energy to the matter. Now we will focus on electrons. It has already been pointed out, that electrons lose their energy in many interactions with the matter. The track of high energy electrons can be quite long, even up to few centimetres. Therefore, it is natural that for electrons the interesting value is the rate of energy loss per unit of length of its path through matter. Because the loss of energy is a stochastic process, we are interested in the expectation value of the energy loss. This value is called stopping power.

$$S = \frac{dE}{dx}\left[\mathrm{MeV/cm}\right] \tag{7}$$

In dosimetry it is better to measure the distance in units related to the density of the absorber. Then the stopping power becomes the mass stopping power.

$$\frac{S}{\rho} = \frac{dE}{\rho \cdot dx}\left[\mathrm{MeV{\cdot}cm^2/g}\right] \tag{8}$$

The stopping power depends on the energy of the electron. The numerical values of stopping powers are rarely measured. Usually these values are calculated according to theory or obtained by means of Monte Carlo methods.

In a similar way as for photons, let us calculate the energy imparted by electrons to a small layer of matter. To simplify matters we shall assume that electrons do not change the direction while travelling across the layer. Then the energy imparted to the layer is:

$$\varepsilon = N \cdot S \cdot z \tag{9}$$

To be precise the energy imparted is slightly smaller because in some interactions the Bremmstrahlung radiation is generated and these Bremmstrahlung photons will not loose their energy in the layer. The absorbed dose can be approximated with the formula:

$$D = \frac{N \cdot S \cdot z}{\rho \cdot A \cdot z} = \frac{N}{A} \cdot \frac{S}{\rho} = \Phi_e \cdot \frac{S}{\rho} \tag{10}$$

To distinguish the electron fluence from the photon fluence the subscript "e" was used. The analogy with the formulae [5] is easily seen.

2.4 Electron Equilibrium

There is one more concept which is important for dosimetry. This is the electronic equilibrium. The electronic equilibrium is defined only for non direct ionizing radiation, i.e. for those particles not bearing a charge. The electronic equilibrium exists in a volume if for each electron of velocity $\overline{v}$ entering the volume there is an identical electron, i.e. of velocity $\overline{v}$ leaving the volume. If the electronic equilibrium exists then Kerma is equal to the absorbed dose. In real life situations the electronic equilibrium does not exist, but in many clinical situations the so called transient electronic equilibrium – Kerma is very close to absorbed dose.

Gases are ionized by ionizing radiation. Let's consider a small volume of air exposed to ionizing radiation. For example, X rays from a point source enter the internal volume of a condenser. In dosimetry such a system is called standard air chamber (in fact the design of a standard air chamber is a little more complex). The air is ionized, electrons are set into motion and attracted by the positively charged wall of the chamber. It might happen that some electrons travelling forward will escape from the air chamber volume but if the electron equilibrium exist then exactly the same number of electrons of the same energy enter the volume. The liberated charge in the volume of a standard air chamber can be measured with an electrometer. It turns out that the average energy required to liberate one pair of ions in air is constant for widely varying conditions of air pressure and electron energies. To produce one pair of ions the energy of 33.97 eV is required. Knowing the mass of air exposed to radiation one may easily calculate Kerma in air.

$$Kerma_{air} = \frac{Q \cdot W/e}{m}, \tag{11}$$

where

Q = measured charge
W/e = 33.97 eV
m = mass of air in the active part of the chamber

As it has already been mentioned in the analyzed situation, electronic equilibrium exists, so the Kerma is very close to the absorbed dose. Thus:

$$D_{air} = Kerma_{air} \cdot (1-g) = \frac{Q \cdot W/e}{m} \cdot (1-g) \tag{12}$$

g is the fraction of electron energy lost to Bremmstrahlung.

In clinical practice we are not interested in absorbed dose to air but we are interested in absorbed dose to tissue. Human soft tissue is mostly composed of water so our task is to know how to measure the dose to water or how to convert the dose measured in air to dose to water.

2.5 Comparison of Absorbed Doses to Two Different Materials

Let us imagine two different materials, 1 and 2. Let us also assume that the electron fluence at the point of interest is the same. We may write:

$$D_1 = \Phi_e \cdot \left(\frac{S}{\rho}\right)_1 \text{ and } D_2 = \Phi_e \cdot \left(\frac{S}{\rho}\right)_2 \tag{13}$$

Then:

$$\frac{D_1}{D_2} = \frac{\Phi_e \cdot \left(\frac{S}{\rho}\right)_1}{\Phi_e \cdot \left(\frac{S}{\rho}\right)_2} = \frac{\left(\frac{S}{\rho}\right)_1}{\left(\frac{S}{\rho}\right)_2} \tag{14}$$

If we can measure the dose in one material we may calculate the dose to the other one with the above formulae. The formula is known as the Bragg-Gray theory. For an air chamber the formulae will assume the following form:

$$D_{water} = D_{Air} \cdot \frac{\left(\frac{S}{\rho}\right)_{water}}{\left(\frac{S}{\rho}\right)_{Air}} \tag{15}$$

Formula 15 is valid only when the electronic equilibrium exists. To establish electronic equilibrium the size of a standard air chamber should be close to the range of electrons liberated by photons. For 3 MeV photons the maximum range of liberated electrons is about 1,500 cm. The construction of such large chambers creates many problems and is very impractical. Therefore, instead of a standard air chamber the thimble chamber has been designed.

3 Dosimeters for Ionizing Radiation Used in Clinical Practice

There are many dosimeters used in clinical practice. The choice of the most appropriate dosimeter in a given situation requires analysis of its characteristics. The perfect dosimeter should at least be accurate, linear, should not have dose rate, energy and directional dependence and should allow the measurement of the dose at a given point (high spatial resolution). There is no one dosimeter that fulfils all these requirements. In the next part of this lecture we will describe several detectors.

3.1 Thimble (Cylindrical) Chamber

It has already been mentioned that the standard air chamber may be used only for relatively low energy. Such low energies are not used in clinical practice today. For high energy photons and “less frequently for electrons” the thimble chamber is used

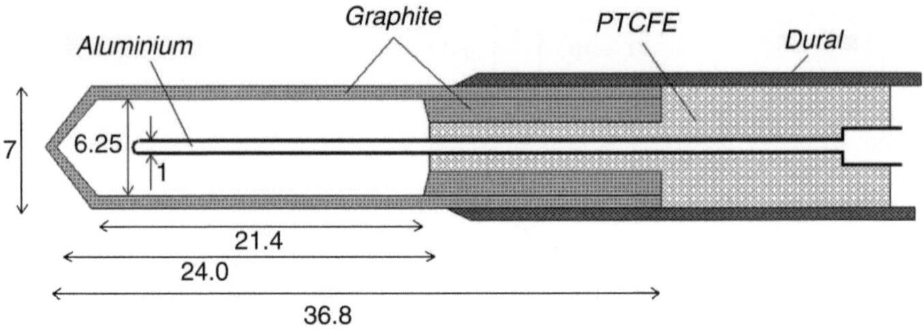

Fig. 3 The farmer chamber

instead of the standard air chamber. Figure 3 shows the most popular thimble chamber, the Framer thimble is shown.

The charge produced by electrons interacting with air is measured (it is the same principle as for the standard air chamber). The electronic equilibrium or more precisely the transient electronic equilibrium is ensured by the chamber wall and the surrounding matter. The dose absorbed in the air of the chamber cavity is converted to absorbed dose at a given point in medium (water) placed in the centre of the chamber. According to the Bragg-Gray theory, the dose to the medium is related to the dose in the cavity with formulae [15]. Because the measurement situation always slightly departs from the Bragg-Gray theory assumptions, the perturbation correction factor p_u is added to formulae [15], which becomes:

$$D_{water} = D_{Air} \cdot \frac{\left(\frac{S}{\rho}\right)_{water}}{\left(\frac{S}{\rho}\right)_{Air}} \cdot p_u$$

There is no room here to explain all perturbation correction factors that have to be applied. Usually the p_u has the form:

$$p_u = p_1 \cdot p_2 \cdot \ldots p_k \tag{16}$$

There is one perturbation correction factor which, from a didactic point of view is very interesting. The wall of the chamber cannot be constructed from water so the material used interacts with the radiation beam in a slightly different way than water. Therefore, the energy which is transferred from the photons to electrons differs a little from the energy transferred in the case of water. This difference must be accounted for. The energy transferred from photons to electrons per unit mass is simply Kerma. We may write:

$$\left(\frac{E_{tr}}{m}\right)_{wall} = K_{wall} = \left(\frac{\mu}{\rho}\right)_{wall} \cdot \left(E_{tr,avr}\right)_{wall} \cdot \Phi \tag{17}$$

$$\left(\frac{E_{tr}}{m}\right)_{water} = K_{water} = \left(\frac{\mu}{\,}\right)_{water} \cdot \left(E_{tr,avr}\right)_{water} \cdot \Phi \tag{18}$$

Subscripts “wall” and “water” denote the material of which the chamber’s wall and the water are made respectively. The ratio of K_{water} and K_{wall} describe the perturbation correction factor for material and is given by (designations are the same as for formulae 5):

$$p_{mat} = \frac{K_{water}}{K_{wall}} = \left[\left(\frac{\mu}{\rho}\right)_{water} \cdot \left(E_{tr,avr}\right)_{water}\right] \Big/ \left[\left(\frac{\mu}{\rho}\right)_{wall} \cdot \left(E_{tr,avr}\right)_{wall}\right] \tag{19}$$

This correction factor should be used to account for different materials of the wall and water (Strictly, the situation is little more complicated because usually only a small part of electrons that ionize the air in the chamber are generated in the chamber’s wall).

The thimble chamber is a very good dosimeter for calibration of the treatment beams. The uncertainty performed with a thimble chamber is small. The reading is proportional to the signal measured. There is a very small energy and dose rate dependence. Most thimble chambers have no directional dependence. Because the size of the chamber is rather large, it is not the well suited for measurements of dose distributions in regions of high dose gradient. For some applications, the so called plane-parallel chamber has been designed. This type of chamber is recommended for measurements of depth dose distributions for photon and electron beams. For brachytherapy sources, the so called well chambers more suitable for the calibration of sources.

3.2 *Films*

Films have very wide applications in radiation dosimetry they can serve as radiation detectors in radiotherapy especially for high gradient dose distributions, as detectors used in radiation protection and as an archival medium. The useful dose range of a film is limited. Films are slightly more sensitive for low energy photons. A proper application of the films is very demanding. The dependence of the film blackness on dose has to be established. The film blackness is expressed in the so called optical density (OD). The OD is read with film densitometers, laser densitometers and film scanners.

There are two types of films used, the radiographic and GafChromic films.

The first one has been used from the beginning of the discovery of X-rays, however at first not for dosimetry purposes. The latter became more and more popular today because it has almost perfect dosimetric features. Its composition resembles closely that of a tissue and it develops a blue colour upon radiation exposure without processing. A little problem is the lack of precise film scanners for GafChromic films.

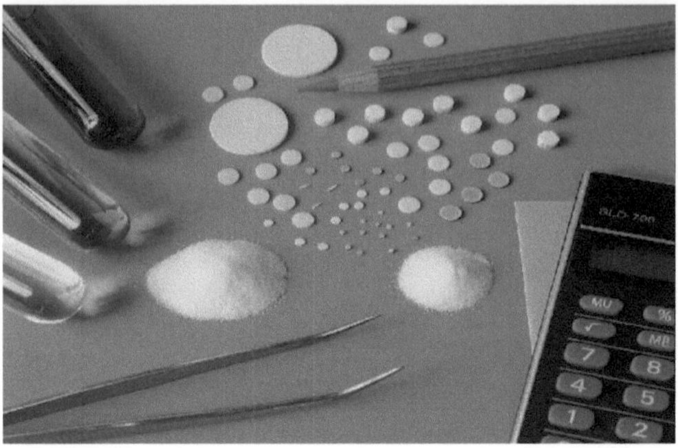

Fig. 4 TLD detectors

In film dosimetry, if proper care is taken of the whole chain of measurements performed with films, a precision better than 3% is achievable.

3.3 Thermoluminescent Dosimetry

During absorption of radiation energy the secondary charged particles move electrons into the conduction band of the TLD material. When the TLD material is heated, the trapped electrons are released together with emitted light that is proportional to the absorbed dose. The thermoluminescent dose response is linear over wide range of doses. Unfortunately, the energy dependence of this type of detectors is pronounced for low energy radiation. Also a correction factor for fading must be applied (correction factor for the time between irradiation and reading). The thermoluminscent detectors have often been used for in-vivo dosimetry, for verification of the accuracy of calculations performed with the treatment planning systems especially for new treatment techniques, and for radiation protection as an individual's personal dosimeter. The most important role of TLDs detectors is to serve as dosimeters for postal dose audit programmes (Fig. 4).

3.4 Diodes

A silicon diode dosimeter is a p-n junction diode. In radiotherapy p type (p-si) are used. in a diode exposed to radiation, an electron-hole is produced. Figure 5 shows typical diodes used in radiotherapy. The minority charge carriers produced in the active part of the detector diffuse into the depleted region. The electrical current generated by radiation is proportional to the radiation dose. Compared to the ionization chamber, the diode has the advantage of a high sensitivity which makes it

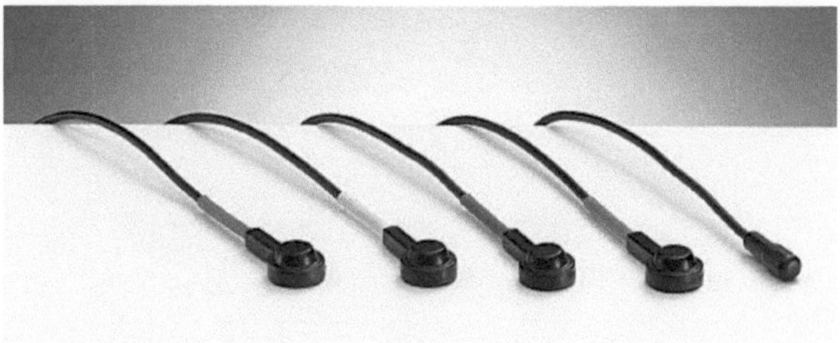

Fig. 5 PTW diodes for in-vivo dosimetry

more reproducible. The active part of the detector is very small which allows for measurements in the area of high gradient dose distribution. The interaction of silicon for high energy radiation (photons and electrons) is very similar to that of water. The energy dependence of TLDs is pronounced for low energy radiation. The diode sensitivity depends on dose rate, and diode temperature. Diodes are widely used for in-vivo dosimetry both in tele- and brachytherapy. Due to their small active part diodes are used for very small fields typical for radiosurgery.

4 Summary

The ability of dose measurement is one of the most important skills of medical physicists. For safety reason the accurate measurement of dose in all application of ionizing radiation for treatment is of utmost importance. Therefore, the understanding of the physical fundamentals of dosimetry plays an important role in the preparation of medical physicists to serve as collaborators of radiotherapists. There are many different detectors developed for dose measurement and none of them serves all purposes. The right decision as to which detector should be used in the actual situation is the responsibility of an expert in medical physics, as he is aware of the characteristics and the limitations of each individual detector.

References

1. E. Podgorsak, Radiation Oncology Physics: A Handbook for Teachers and Students, IAEA, Vienna, Austria, 2005. www.pub.**iaea**.org/MTCD/publications/PDF/Pub1196_web.pdf.
2. H.E. Johns, J.R.Cunningham, The Physics of Radiology, fourth edition, Charles C Thomas Publisher, Springfield, IL, 1983.
3. F. Attix, Introduction to Radiological Physics & Radiation Dosimetry, Wiley-Interscience, New York, 1986.

Dose Assessment (for Patients)

Cornelius Lewis

1 Introduction

One of the requirements of the Medical Exposures Directive is that information should be available from each medical exposure undertaken to enable the assessment of dose to the patient. Normally this information is collected but only rarely is it required for a full dose assessment. Assessments are usually required if there has been an inadvertent exposure, e.g. the wrong diagnostic exposure was performed or an exposure was performed on the wrong patient, or if the patient has been exposed to a particularly high dose in, for example, a lengthy interventional procedure.

2 Key Parameters for Dose Assessment

2.1 Plain Radiography and Fluoroscopy

The quantities normally used for dose assessment in these modalities are Dose Area Product (DAP) and Entrance Surface Dose (ESD) although an alternative would be tube voltage (kV) and current-time product (mAs).

DAP devices are fitted as standard to most modern x-ray units and in some older units they have been retro fitted. The DAP meter is a parallel plate ionisation chamber located in the tube housing or just beneath it. DAP meters are calibrated at commissioning and during regular quality assurance testing to provide a measurement of the product of air kerma and field size (units are $mGy.cm^2$). In essence they provide a measure of air kerma for the particular patient exposure. DAP measurements would normally be recorded for each patient exposure.

C. Lewis (✉)
Kings College Hospital, London, UK
e-mail: Cornelius.Lewis@kcl.ac.uk

Y. Lemoigne and A. Caner (eds.), *Radiation Protection in Medical Physics*,
NATO Science for Peace and Security Series B: Physics and Biophysics 1,
DOI 10.1007/978-94-007-0247-9_10,

Entrance surface dose is a measure of dose at the patient's skin and would include a back scatter factor. ESD is not normally recorded as this would interfere with the examination. However, for most exposures, tables of ESD could be generated.

Tube voltage and (post exposure) mAs is usually recorded for each exposure and this would enable ESD to be determined.

2.2 *Computed Tomography*

The two key parameters for assessing patient dose in CT are the Computed Tomography Dose Index (CTDI) and Dose Length Product (DLP).

CTDI is a derived measurement to enable the assessment of dose from a single CT slice. It is defined as the dose measured in a 100 mm pencil chamber divided by the scan thickness. Thus;

$$CTDI_{100} = \frac{1}{T}\int_{-50}^{+50} D(z)dz$$

where T is the scan thickness and the integral is the dose measured in the pencil chamber.

$CTDI_{100}$ measurements are made during commissioning of the CT scanner and checked at regular quality assurance testing. To use in patient dose assessment the parameter required is the weighted $CTDI_{100}$ measured in a phantom. This parameter is referred to as $CTDI_w$ or, strictly, $CTDI_{100,w}$. Measurements of $CTDI_{100}$ are made at the centre and periphery of a phantom (representing the head or body) and a weighted average taken. Thus,

$$CTDI_{100,w} = \frac{1}{3}CTDI_{100,C} + \frac{2}{3}CTDI_{100,P}$$

where $CTDI_{100,C}$ is the $CTDI_{100}$ measured at the centre of the phantom and $CTDI_{100,P}$ is the average of four $CTDI_{100}$ measurements taken at the periphery of the phantom (3, 6, 9 and 12 o'clock positions).

The dose to a patient from a single slice is thus $CTDI_{100,w}$. In assessments of patient dose this is more commonly referred to as $CTDI_{vol}$. However as most modern CT scanners operate in a helical (spiral) mode the $CTDI_{vol}$ parameter must be modified to take spiral pitch into account. Thus;

$$CTDI_{vol} = \frac{CTDI_w}{pitch}$$

For a pitch of 1;	$CTDI_{vol} = CTDI_w$
For a pitch of 2;	$CTDI_{vol} = CTDI_w/2$
For a pitch of 0.5;	$CTDI_{vol} = 2 \times CTDI_w$

The other useful parameter for patient dose assessment in CT is Dose Length Product (DLP) which is calculated as:

$$DLP = CTDI_{vol} \times L$$

where L is the scan length, usually recorded in cm.

In modern CT scanners one or both of these parameters ($CTDI_{vol}$ measured in mGy or DLP measured in mGy.cm) is displayed on the CT console.

3 What Is Required in a Dose Assessment?

If a dose assessment is requested it will be to determine the risk of damage to specific organs or to determine overall risk to the patient. For example a patient undergoing angioplasty for multiple vessel disease may have been subject to a significant skin dose. In this case there would be concern about deterministic effects to the skin and for this an absorbed dose to the skin would be calculated.

Alternatively, if the wrong patient received an abdominal CT scan there might be concern about the overall risk of subsequent stochastic effects (cancers) resulting from the exposure. Stochastic risk is conventionally related to effective dose using the 5% per Sievert risk factor and subsequently relating to 'background equivalent' (i.e. the number of months/years to acquire the same dose from natural background radiation) or some other risk comparator, e.g. miles driven on the road, air miles travelled, cigarettes smoked, etc. It is also possible, and some argue preferable, to quote individual organ risk factors.

4 Dose Assessment in Plain Radiography and Fluoroscopy

There have been a number of methods proposed for dose assessment in these modalities. At one extreme, simulated exposures in anthropomorphic phantoms can be made with doses measured using TLD. Less precise estimates can also be made using a variety of exposure charts for various procedures. However these are either very time consuming or imprecise.

As the power of personal computers has increased it has become possible to run mathematical simulations using Monte Carlo techniques. These make dose assessment easily accessible and rapid.

The most widely used Monte Carlo technique is a programme known as PCXMC. This has been developed by STUK, the Finnish Radiation and Nuclear Regulatory Authority. The software is provided at a modest cost and details can be found on the STUK website, http://www.stuk.fi.

PCXMC calculates dose and risk estimates from 29 organs and tissues in adult and paediatric mathematical phantoms. It also assesses effective dose using both ICRP 60 and ICRP 103 tissue weighting factors.

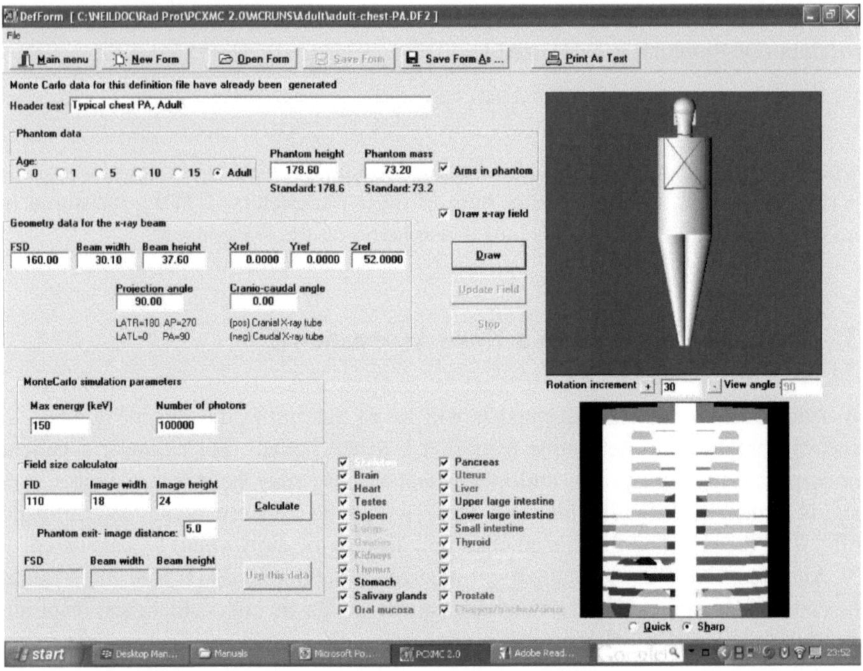

Fig. 1 Main page of PCXMC

The diagram above (Fig. 1) shows the main page of PCXMC where information about particular exposures can be entered into the programme. The example shown above is for a standard chest x-ray.

The data entered to describe the examination includes;

- Patient height
- Patient weight
- Age (adult, or age 15, 10, 5, 1, 0)
- Focus to Skin Distance (FSD)
- Beam dimensions at the skin entrance level
- Projection (AP/LAT and Cranio/Caudal tilts)
- Maximum photon energy

The program includes a calculator to determine beam entrance dimensions if only the FID (Focus to Image Distance) and image size are known. Also there is a display of the projected beam on the exterior of a phantom and also a pseudo radiograph to show the organs irradiated in the exposure. The programme also requires the user to enter the number of photons to be simulated by the Monte Carlo process. With current PC processing speeds, 100,000 photons can be simulated in a matter of seconds. The number of photons simulated will determine the accuracy of the dose estimate.

Once the examination data has been entered the simulation is run and a results file stored. Following this the x-ray spectral data is entered in the form of a maximum kilovoltage (kVp), anode angle and filtration. It is not uncommon for the anode angle to be unknown. Normally this is in the range 11–13° but actually makes only a minor difference to the results. The final data entry requirement is the measurement of exposure which may be given as one of the following;

- Incident air kerma (mGy)
- Dose area product (DAP) (mGy.cm^2)
- Entrance exposure (mR)
- Exposure-area product (R.cm^2)
- Current-time product (mAs)

The program then combines the data stored from the Monte Carlo simulation with the beam quality data and exposure factor to produce a series of individual organ doses and estimates of effective dose using both ICRP 60 and ICRP 103 tissue weighting factors (Fig. 2).

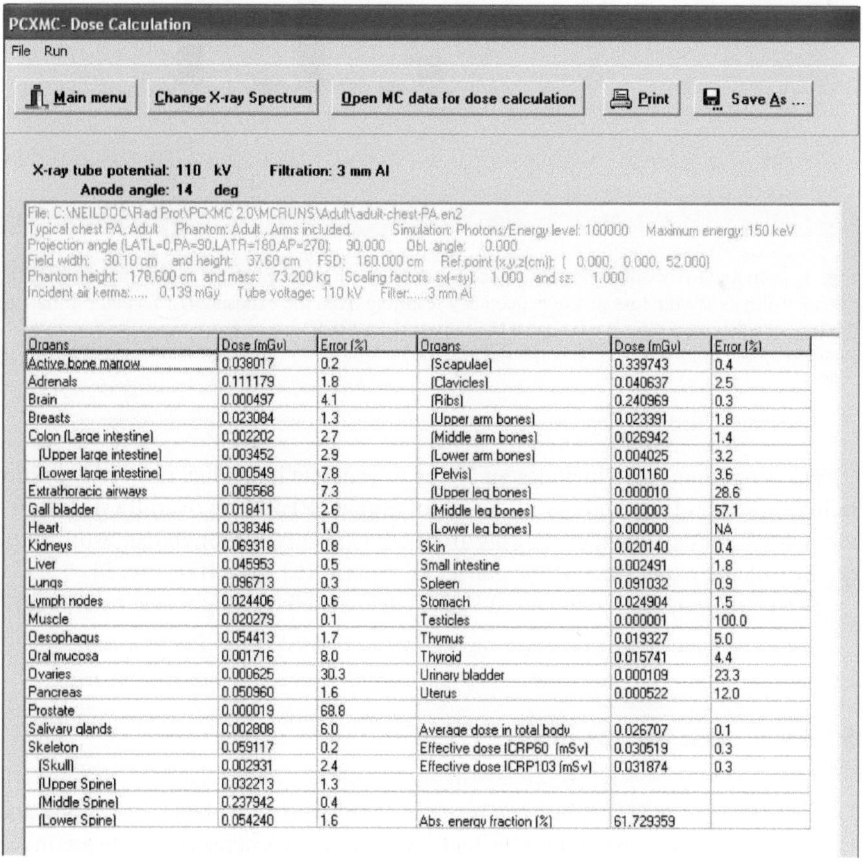

Organs	Dose (mGy)	Error (%)	Organs	Dose (mGy)	Error (%)
Active bone marrow	0.038017	0.2	(Scapulae)	0.339743	0.4
Adrenals	0.111179	1.8	(Clavicles)	0.040637	2.5
Brain	0.000497	4.1	(Ribs)	0.240969	0.3
Breasts	0.023084	1.3	(Upper arm bones)	0.023391	1.8
Colon (Large intestine)	0.002202	2.7	(Middle arm bones)	0.026942	1.4
(Upper large intestine)	0.003452	2.9	(Lower arm bones)	0.004025	3.2
(Lower large intestine)	0.000549	7.8	(Pelvis)	0.001160	3.6
Extrathoracic airways	0.005568	7.3	(Upper leg bones)	0.000010	28.6
Gall bladder	0.018411	2.6	(Middle leg bones)	0.000003	57.1
Heart	0.038346	1.0	(Lower leg bones)	0.000000	NA
Kidneys	0.069318	0.8	Skin	0.020140	0.4
Liver	0.045953	0.5	Small intestine	0.002491	1.8
Lungs	0.096713	0.3	Spleen	0.091032	0.9
Lymph nodes	0.024406	0.6	Stomach	0.024904	1.5
Muscle	0.020279	0.1	Testicles	0.000001	100.0
Oesophagus	0.054413	1.7	Thymus	0.019327	5.0
Oral mucosa	0.001716	8.0	Thyroid	0.015741	4.4
Ovaries	0.000625	30.3	Urinary bladder	0.000109	23.3
Pancreas	0.050960	1.6	Uterus	0.000522	12.0
Prostate	0.000019	68.8			
Salivary glands	0.002808	6.0	Average dose in total body	0.026707	0.1
Skeleton	0.059117	0.2	Effective dose ICRP60 (mSv)	0.030519	0.3
(Skull)	0.002931	2.4	Effective dose ICRP103 (mSv)	0.031874	0.3
(Upper Spine)	0.032213	1.3			
(Middle Spine)	0.237942	0.4			
(Lower Spine)	0.054240	1.6	Abs. energy fraction (%)	61.729359	

Fig. 2 Individual organ doses and estimates of effective dose obtained by using both ICRP 60 and ICRP 103 tissue weighting factors

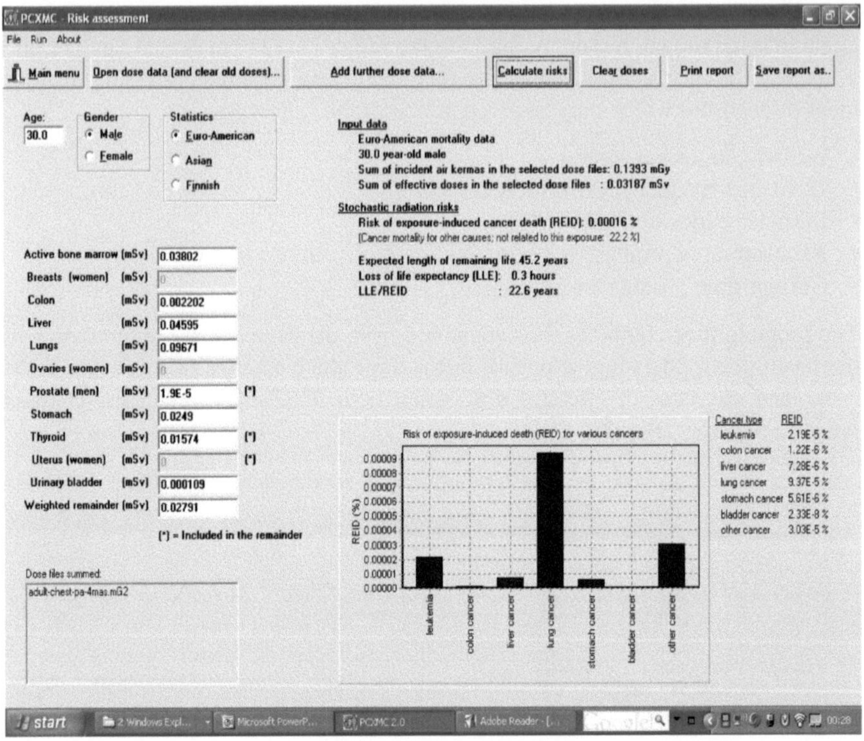

Fig. 3 Individual risks of exposure induced cancers for a number of the major organs (REID), overall estimate for the loss of life expectancy resulting from the exposure (LLE) and estimate of the loss of life expectancy if the cancer is realised (LLE/REID)

The programme also includes an option to generate individual risks of exposure induced cancers for a number of the major organs (REID), an overall estimate for the loss of life expectancy resulting from the exposure (LLE) and an estimate of the loss of life expectancy if the cancer is realised (LLE/REID) (Fig. 3).

These data can also be modified by sex, age and ethnic origin (see over page).

5 Dose Assessment in CT

Dose assessment in CT is, in some respects, simpler than in plain radiography and fluoroscopy because the conditions under which the exposures are made are much more tightly controlled. Also, with modern CT scanners the dose information required to make the assessment is displayed at the computer console.

Table 1 DLP conversion factors

Region of body	E/DLP conversion factor (mSv/mGy.cm)
Head	0.0023
Neck	0.0054
Chest	0.019
Abdomen	0.017
Pelvis	0.017
Legs	0.008

At the simplest level, because of the fixed geometry of CT scans, it is possible to relate the Dose Length Product (DLP) to effective dose for scans of specific body areas. The table that relates DLP to effective dose is shown below (Table 1).

To estimate individual organ doses requires a more detailed approach. The UK National Radiological Protection Board (now the Radiological Protection Division of the Health Protection Agency) undertook a series of Monte Carlo calculations to determine individual organ dose for 22 organs from a range of CT scanners that were in use at the time. These data, known as NRPB SR250, which comprise 23 data sets are available at a small cost from the UK Health Protection Agency (http://www.hpa.org.uk).

The ImPACT CT scanner assessment centre, based at St Georges Hospital in London, has developed an Excel spreadsheet which enables the NRPB data to be used to calculate individual organ doses. This is available as a free download from their website, http://www.impactscan.org. The ImPACT spreadsheet is regularly updated to match current CT scanners with the original data sets produced by NRPB. The spreadsheet is currently at Version 0.99x.

The following extract from the spreadsheet indicates the data required for a dose assessment (Table 2).

The data entry and results screen for the spreadsheet is shown on the following page together with the screen for selecting body region from the mathematical phantom.

Whilst this is a very powerful and useful piece of software it must be used intelligently. For example, in almost all cases the length of the scan recorded for a particular patient will not map directly onto the standardized mathematical phantom. The scan length entered into the software should reflect the range of organs irradiated rather than the total length of the scan. This is particularly relevant for scans acquired with gantry angulation

There are also modifications which should be made to accommodate 'overscanning' in spiral acquisitions. A good approximation is to extend the scan by the collimated width of the beam at either end of the acquisition.

Finally, the data can be modified to correct for age using estimated correction factors. The table of correction factors, which may be as much as an increase of 2.4 for abdomen scans in infants, is also included as a page in the spreadsheet (Figs. 4 and 5).

Table 2 Spreadsheet for dose assessment

Using CTDosimetry.xls

To calculate doses using CTDosimetry.xls, the user must enter a number of parameters relating to the scanner and the scan series

The following four selections, made in the top left box on the ScanCalculations worksheet define the Monte Carlo data set that is used:

Manufacturer	Select the scanner manufacturer from the drop down list
Scanner	Select the scanner model or scanner model group for the drop down list
kV	Choose the appropriate scan kV
Scan region	Choose head or body

The Monte Carlo data set that is used for this combination of scanner, kV and body part is displayed in the cell marked 'Data Set'. The data set that is currently loaded is displayed below. If these do not match, no dose is calculated. To load the appropriate data set, and enable dose calculation, press the 'Update Data Set' button

Scan and patient data is entered in the box on the top right of the ScanCalculations worksheet

mA	The x-ray tube current. Note that this should be the actual scanner mA, and not the 'effective mAs' displayed on some multi-slice scanners
Rotation time	The scanner tube rotation time
mAs/rotation	Do not enter data in this box (it is calculated automatically)
Collimation	The total nominal x-ray beam width along the z-axis, selected from a range of possible values in the drop down box. This determines the relative CTDI compared to the reference (usually 10 mm) collimation
Slice width	The scanner collimation slice width. This is not actually used in calculations, but can be useful in printed output
Pitch	The scanning pitch (table travel per rotation/total collimated slice width). For axial scanning (couch increment)/(collimated slice width) should be used
Rel. CTDI	The CTDI at the selected collimated x-ray beam thickness, relative to the CTDI at the reference collimation (usually 10 mm)
CTDI (air)	The free in air $CTDI_{100}$ value (in mGy/100 mAs), as defined in EUR 16262: European Guidelines on Quality Criteria for Computed Tomography, pub. European Commission (link to this document at bottom of page). CTDI values for most of the scanners are listed on the Scanner Worksheet. Pressing the 'Look up' button will enter the value in this cell. The value in this cell is corrected for the relative CTDI value in the cell above
CTDI (soft tissue)	The CTDI to ICRU muscle, used as an approximation to the dose to soft tissue within the body. This is the CTDI(air) × 1.07 for CT scanner energies
$_{n}CTDI_{w}$	Weighted $CTDI_w$ measured in a standard CTDI phantom (normalised for 100 mAs). $CTDI_w = (CTDI_{centre} + 2 \times CTDI_{periphery})/3$. See EUR16262 for more details (link below)
Patient Sex	Enter 'm' or 'f' in this cell for male or female patients. This affects the organ used for gonad dose calculation. If left blank, the program will use an average value
Start position	The start position of the scan series. The diagram on the Phantom worksheet shows the position of the phantom's organs relative to the number scale, which is zero at the base of the trunk. This value can be entered manually in the worksheet, or can be taken from the shaded area on the Phantom worksheet diagram. This can be adjusted using the up and down arrows. Pressing the 'Get From Phantom Diagram' button enters these values into the start and end position boxes in ScanCalculation

(continued)

Table 2 (continued)

End position	The end position of the scan series – Note that this should include the slice thickness, so, for example, a single 5 mm slice 20 cm from the base of the trunk would have a start position of 20, and an end position of 20.5 cm. Start and End position values are interchangeable

When the above values are entered, the doses to each of the individual organs, as defined by the SR250 data set appear in the cells below the scan parameters. These are combined according to the tissue weighting factors given in ICRP60, to calculate an effective dose

In addition, the weighted CTDI (CTIDw), volume CTDIw (CTDIvol) and dose length product (DLP) are also displayed

Note that not all of the ICRP60 organs are included in NRPB SR250. In order to estimate dose to the oesophagus, the thymus dose is used. The dose for muscle is approximated from the total body dose – dose to all other organs and contents

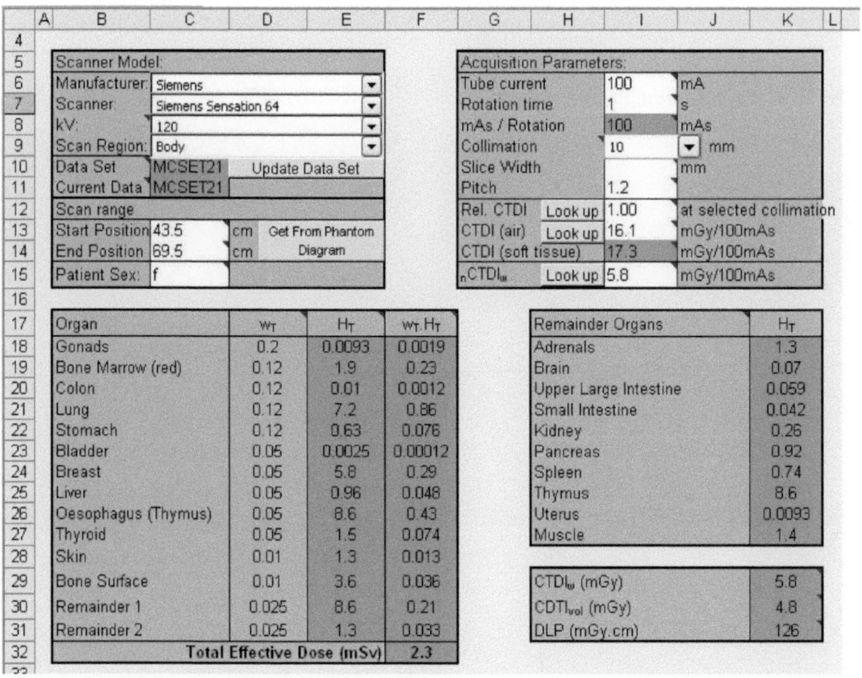

Fig. 4 Spreadsheet data entry and results screen

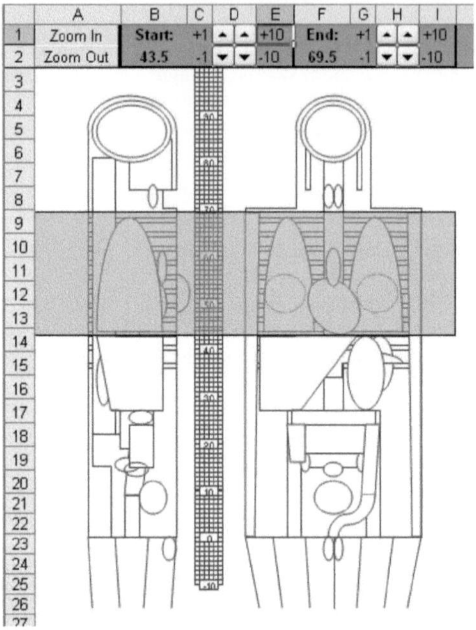

Fig. 5 Screen for selecting body regions

Patient Dose Assessment in Nuclear Medicine

Stelios Christofides

Abstract The ultimate goal of any type of medical imaging procedure is to obtain the best image quality while delivering the smallest radiation dose possible to the patient. The best image quality though, does not necessarily give the correct diagnosis for a given medical condition at the lowest possible dose to the patient. Additionally the vast number of alternative diagnostic modalities available today and their rapid evolution make the choice of the most suitable modality for a particular medical condition very difficult, if dose to the patient is to be considered as a major constraint. It is therefore very important to know the dose received by the patient from the different modalities to arrive at the same diagnostic result. This is especially important in Nuclear Medicine where the different modalities produce images of the metabolic function of the human body and they are more likely to arrive at the same diagnostic outcome. The aim of this presentation is to give an overview of the peculiarities of the management of patient doses in Nuclear Medicine diagnostic procedures and to present a review of the methods used to estimate organ doses from the various radiopharmaceuticals used in diagnostic Nuclear Medicine procedures.

Keywords Nuclear medicine • Organ dose • Phantoms • Radionuclides

1 Introduction

The diagnostic value of each radiopharmaceutical is given by its specificity in the organ being examined and the sensitivity of the modality detecting the radiation emitted by the radiopharmaceutical. The Quality of the image produce depends, among other parameters, on the amount of activity given to the patient as well as the time taken to acquire enough counts to build up the image. Too much activity

S. Christofides (✉)
Medical Physics Department, Nicosia General Hospital, Strovolos, Nicosia, Cyprus
e-mail: cstelios@cytanet.com.cy

Y. Lemoigne and A. Caner (eds.), *Radiation Protection in Medical Physics*,
NATO Science for Peace and Security Series B: Physics and Biophysics 1,
DOI 10.1007/978-94-007-0247-9_11,

gives images with a high noise background that may hide valuable information. Not enough activity will require more time to acquire enough counts to build up the image, with the danger of patient movements. These will produce movement artefacts in the image that again will reduce the diagnostic value of the image.

It is therefore important to know the dose received by the different body organs so that this will not exceed the recommended diagnostic reference levels and therefore increase the risk to the patient.

The organ dose calculation is a complicated issue that requires the detailed study of the bio-kinetic behaviour of the radiopharmaceutical of interest. Through the years a lot of work has been done in this area using different techniques and models in order to arrive at established data for existing and well-established radiopharmaceuticals. These will be briefly reviewed in this presentation.

2 Metabolic Imaging

The Nuclear Medicine diagnostic procedures study the metabolic function of the human body and therefore the human body must be alive in order to undergo a Nuclear Medicine examination. This is achieved by the administration of a radiopharmaceutical that it is designed to concentrate in the particular organ of interest. This is traced or imaged by the appropriate Nuclear Medicine instrumentation in order to collect the necessary data to produce an image, a series of images or other dynamic metabolic function curves that will assist the Nuclear Medicine Physician to make a diagnostic evaluation of the patient. The factors affecting the exposure to the patient are:

1. The Equipment and other instrument used.
2. The Pharmaceutical used.
3. The procedure used.
4. The patient Physiology.

In this presentation 2 and 4 will only be elaborated further since these the factors directly related to the organ dose received by the administered activity.

2.1 Radiopharmaceuticals

Different radiopharmaceuticals are used depending on the Molecular Imaging modality used (PET or SPECT). Also for a specific examination there may be more that one radiopharmaceutical that can be used to acquire the final image. Table 1 lists examples of radiopharmaceuticals used with Gamma Camera systems and Table 2 lists examples of Radionuclides used with PET systems.

Table 1 Examples of radiopharmaceuticals used with gamma camera systems

Tc-99m Albumin microspheres	Tc-99m Sulfur colloid (liver disease)
Tc-99m DMSA	Ga-67 Citrate
Tc-99m DTPA aerosol	In-111 Platelets
Tc-99m DTPA injection	In-111 RBC
Tc-99m HMPAO	In-111 WBC
Tc-99m MAA	I-123 NaI
Tc-99m MAG3	I-123 Hippuran
Tc-99m MDP	I-123 mIBG
Tc-99m MIBI	I-125 NaI
Tc-99m Pertechnetate injection	I-125 mIBG
Tc-99m WBC	I-131 NaI
Tc-99m Pertechnetate infusion	I-131 Hippuran
Tc-99m RBC (in-vivo labeling)	I-131 mIBG
Tc-99m RBC (in-vitro labelling)	Tl-201 Chloride

Table 2 Examples of radionuclides used with PET systems

Radionuclide	Half-life (min)	Mean particle energy (MeV)
C-11	20.40	0.39
N-13	10.00	0.50
O-15	2.20	0.72
F-18	110.00	0.25
Cu-62	9.20	1.30
Ga-68	68.30	0.83
Rb-82	1.25	1.50

The diagnostic value of each radiopharmaceutical is given by its specificity in the organ being examined and the sensitivity of the modality detecting the radiation emitted by the radiopharmaceutical.

Also the maximum amount administered should not exceed the maximum allowed organ dose. As the injected radiopharmaceutical circulates in the blood system before it is absorbed and preferentially concentrate in the organ of interest, other organs of the body absorb some of the radiopharmaceutical and therefore receive a dose proportional to the amount of radiopharmaceutical absorbed, taking into account the composition of the organ and the type of radiopharmaceutical.

2.2 *Patient Physiology*

In order to maintain the dose to the patient as low as reasonable and at the same time receiving the best diagnostic value from a Nuclear Medicine Examination, it is necessary to administer the correct amount of radiopharmaceutical to the patient. This amount depends on the following patient parameters that are more important for paediatric patients:

1. Anatomy.
   - Body proportions.
   - Size and shape of Organs.
   - Skeletal Development.
   - Bone marrow distribution shift during growth.
2. Pathology.
   - Specific age dependant diseases.
3. Biochemistry.
   - Total body water (newborns 90%, adults 60%).
   - Fat mass.
   - Different Metabolism.
4. Physiology.
   - Respiratory rate (newborns 80–120 breaths/min).
   - Heart rate (newborns 160–200 beats/min).
   - Body movements (fast movements of paediatric patients).

Therefore the amount of administered radiopharmaceutical depends on the patient's:

5. Gender.
6. Age.
7. Height.
8. Weight.

Details of how some the above patient factors affect the amount of administered radiopharmaceutical is given below under the section on organ dose calculations.

The required activity to be administered is given by the manufacturer of the radiopharmaceutical in the leaflet accompanying the radiopharmaceutical. This amount always refers to the standard 70 kg man.

Table 3 gives the fraction of adult administrated activity for different age groups of children based on weight.

3 Organ Dose Calculations

The organ dose calculation is a complicated issue that requires the detailed study of the bio-kinetic behaviour of the radiopharmaceutical of interest. Through the years a lot of work has been done in this area using different techniques and models in order to arrive at established data for existing and well-established radiopharmaceuticals. These are given in table format or calculated by the use of computer software taking into account all the factors mentioned previously.

Table 3 Fraction of adult administrated activity for different age groups of children based on weight

Weight (kg)	Fraction of adult administrated activity	Weight (kg)	Fraction of adult administrated activity	Weight (kg)	Fraction of adult administrated activity
3	0.1	22	0.50	42	0.78
4	0.14	24	0.53	44	0.80
6	0.19	26	0.56	46	0.82
8	0.23	28	0.58	48	0.85
10	0.27	30	0.62	50	0.88
12	0.32	32	0.65	52–54	0.90
14	0.36	34	0.68	56–58	0.95
16	0.40	36	0.71	60–62	1.00
18	0.44	38	0.73	64–66	
20	0.46	40	0.76	68	

The resulting organ dose depends on:

- The physical properties of the radionuclide
- The anatomical properties of the patient
- The biokinetics of the radiopharmaceutical in the patient

The physical properties comprise the radiation type and energy as well as the physical half-life of the particular radionuclide involved. The essential anatomical properties are the mass, shape and arrangement of the organ in the body. Biokinetics describes the time-dependent distribution of radioactivity in individual organs or in the whole body. Anatomy and biokinetics are also dependent on the age and functional status of the body.

3.1 Evaluation Methods

Through the years a lot of work has been done in this area using different techniques and models in order to arrive at established data for existing and well established radiopharmaceuticals. These are given in table format or calculated by the use of computer software taking in to account all the factors mentioned previously.

3.1.1 Compartmental Analysis Method

The compartmental analysis method will be demonstrated in order that the complexity involved will be appreciated.

The organs of the body are considered to be interconnecting compartments through which the radiopharmaceutical passage is constrained according to the metabolic function of the particular organ. The open two compartments in series system is illustrated in Fig. 1.

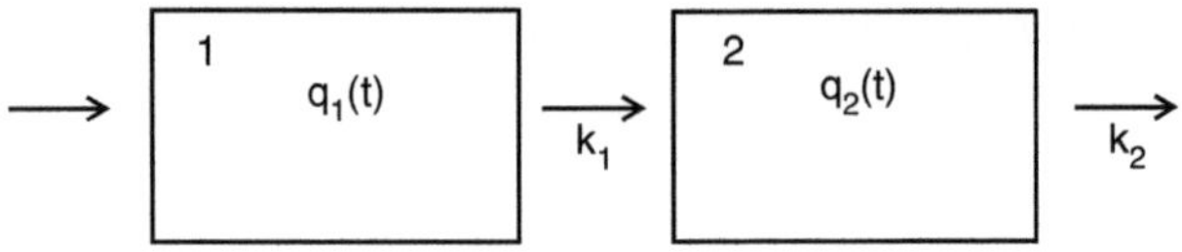

Fig. 1 The open two compartments in series system

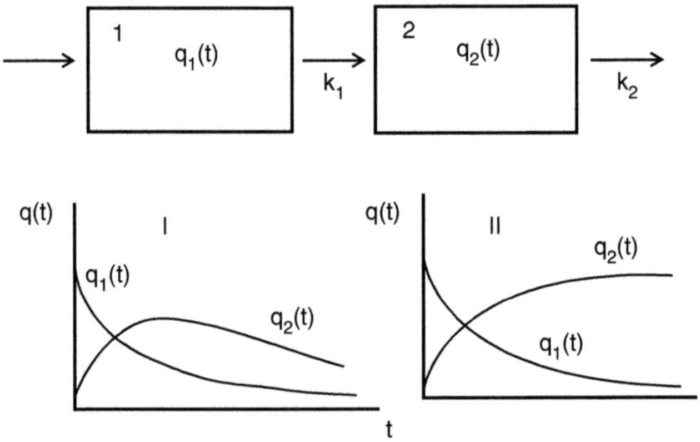

Fig. 2 Curves showing the activity in each compartment over time. In the case 1 when k2≠0 and in case 2 when k2=0

The equations for the variations of the radiopharmaceutical concentrations in compartment 1 and 2 over time are given by:

$$\frac{dq_1}{dt} = -k_1 q_1 \tag{1}$$

$$\frac{dq_2}{dt} = k_1 q_1 - k_2 q_2 \tag{2}$$

Integration of the first equation is immediate

$$q_1(t) = q_0 e^{-k_1 t} \tag{3}$$

Integration of Eq. 2 gives

$$q_2(t) = q_0 \frac{k_1}{k_2 - k_1}(e^{-k_1 t} - e^{-k_2 t}) \tag{4}$$

Graphs of Eqs. 3 and 4 are shown in the next figure for the general case of k2≠0 (plot I) and k2=0 (plot II) (Fig. 2).

The cumulated activity in each compartment is given by the area under the curve. This is essentially a measure of the total number of disintegrations occurring during the time that radioactivity is present in the source organ.

The energy from the disintegrations is absorbed by the source organ as well as the surrounding organs (target organs). The absorbed energy depends on the physical characteristics of the organs as well as the type and energy of the emitted radiation.

If we consider the entire human body as interconnecting compartments the problem becomes very complicated. To calculate the absorbed dose to a particular organ we will need a lot of computational power to solve all the complicated integration equations involved.

A number of mathematical models have been develop that represent the human body (male and female) at the different stages of development.

It is the responsibility of the radiopharmaceutical manufacturers to provide the organ doses per unit of administered radiopharmaceutical on the instructions for use (IFU) leaflet of their products. Examples of such tables are given in Tables 4 and 5.

3.1.2 Tracer Kinetic Modelling

To aid in the solution of the organ dose calculation problem several phantom models have been developed that represent the human body in Monte Carlo calculations (e.g., [1–5]).

These include voxel-phantoms, which are based on computer tomography (CT) and MR images of actual human beings, and computational models where body contours and organs are defined by mathematical expressions. As an example the simplified kinetic model for FDOPA in the striatum is illustrated in Fig. 3.

3.1.3 Phantoms and Mathematical Models

In mathematical phantoms the size and shape of the body and its organs are described by mathematical expressions representing combinations and intersections

Table 4 Radiation dose estimates for Ga-67 Citrate (Estimated radiation dose equivalent in mSv/MBq)

Organ	Newborn	1-year-Old	5-year-Old	10-year-Old	15-year-Old	Adult
Adrenals	0.62	0.56	0.36	0.26	0.18	0.13
LLI wall	1.1	0.54	0.61	0.43	0.26	0.20
Small intestine	0.66	0.30	0.21	0.15	0.095	0.076
ULI wall	0.78	0.36	0.35	0.24	0.15	0.12
Kidneys	1.2	0.48	0.28	0.19	0.14	0.11
Liver	1.3	0.58	0.32	0.22	0.15	0.11
Ovaries	0.84	0.37	0.22	0.15	0.099	0.079
Bone surfaces	17.0	5.3	2.3	1.4	0.84	0.65
Red marrow	4.2	1.4	0.68	0.38	0.23	0.21
Spleen	2.1	0.83	0.46	0.30	0.20	0.14
Testes	0.70	0.29	0.16	0.10	0.064	0.052
EDE	1.8	0.69	0.38	0.24	0.15	0.12

Table 5 Radiation dose estimates for Tc-99m DMSA (Estimated radiation dose equivalent in mGy/MBq)

Organ	1-year-old	5-year-old	10-year-old	15-year-old	Adult
Adrenals	6.0 E-02	3.5 E-02	2.4 E-02	1.6 E-02	1.3 E-02
Bladder wall	9.4 E-02	5.1 E-02	3.5 E-02	2.4 E-02	1.9 E-02
Bone surfaces	1.9 E-02	9.9 E-03	6.4 E-03	4.3 E-03	3.5 E-03
Breast	8.4 E-03	4.5 E-03	2.8 E-03	1.8 E-03	1.8 E-03
Stomach wall	2.0 E-02	1.3 E-02	9.8 E-03	6.3 E-03	5.5 E-03
Small intestine	2.5 E-02	1.5 E-02	1.0 E-02	6.4 E-03	5.2 E-03
ULI wall	2.3 E-02	1.4 E-02	9.6 E-03	6.3 E-03	5.1 E-03
LLI wall	1.8 E-02	1.0 E-02	6.7 E-03	4.2 E-03	3.2 E-03
Kidneys	7.3 E-01	4.2 E-01	2.9 E-01	2.1 E-01	1.7 E-01
Liver	4.1 E-02	2.5 E-02	1.8 E-02	1.2 E-02	9.7 E-03
Lungs	1.4 E-02	8.0 E-03	5.2 E-03	3.5 E-03	2.5 E-03
Ovaries	2.0 E-02	1.1 E-02	7.2 E-03	4.6 E-03	3.7 E-03
Pancreas	3.7 E-02	2.3 E-02	1.6 E-02	1.1 E-02	9.0 E-03
Red marrow	2.0 E-02	1.4 E-02	1.0 E-02	7.5 E-03	6.3 E-03
Spleen	6.1 E-02	3.8 E-02	2.6 E-02	1.7 E-02	1.3 E-02
Testes	1.2 E-02	6.2 E-03	3.9 E-03	2.4 E-03	1.8 E-03
Thyroid	9.2 E-03	5.1 E-03	3.1 E-03	1.9 E-03	1.1 E-03
Uterus	2.3 E-02	2.3 E-02	8.9 E-03	5.5 E-03	4.6 E-03
Other tissue	1.4 E-02	8.0 E-02	5.2 E-03	3.6 E-03	3.0 E-03
EDE	6.9 E-02	4.0 E-02	2.7 E-02	1.9 E-02	1.6 E-02

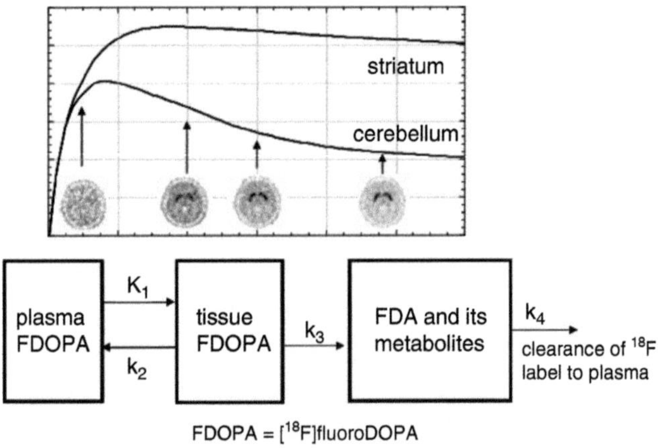

Fig. 3 An example of a simplified model for FDOPA kinetics in Striatum. The rate constants K1, k2, k3 & k4 can be estimated using measured PET time activity curves and blood input function

of planes, circular and elliptical cylinders spheres, cones, etc. Examples are the National Radiological Protection Board (NRPB) Mathematical Phantom, the Medical Internal Radiation Dose Committee, pamphlet no 5 (MIRD5) Phantom and the Oak Ridge National Laboratory (ORNL) Phantom. An example of how MIRD5 describes the brain is shown in Fig. 4.

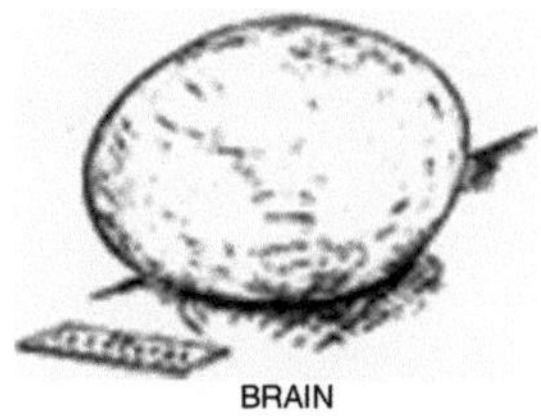

The brain is an ellipsoid given by

$$\left(\frac{x}{6}\right)^2 + \left(\frac{y}{9}\right)^2 + \left(\frac{z-86.5}{6.5}\right)^2 \leqq 1,$$

and the volume is 1,470 cm^2

Fig. 4 An example of how MIRD5 describes the brain

Voxel phantoms are based on digital images recorded from scanning of real persons by computer tomography (CT) or magnetic resonance imaging (MRI). Examples are the Gibbs Phantom (1984), the NORMAN Phantom which is based on MRI data of a volunteer and the Zubal Phantom based on CT and MRI data.

3.1.4 Tracer Kinetic Model Fitting Program

The Tracer Kinetic Model Fitting (TKMF) Program is a server-based tracer kinetic analysis software program that is accessible through the Internet (http://dragon.nuc.ucla.edu/modelfitting/).

It provides the general capability of tracer kinetic model fitting to PET/SPECT generated kinetics that characterise local tissue functions in nuclear medicine.

With simple user interactions, the program automatically fits a set of kinetic tissue data (e.g., measured with PET/SPECT and obtained with ROI analysis) with a user selected tracer kinetic model.

4 Concluding Remarks

In Nuclear Medicine the dose to the organs received by the administered radiopharmaceutical cannot be measured directly. It can only be estimated. These estimations are based on animal experiments, mathematical models and phantoms that help to simulate tracer kinetics.

Today it is possible to use real patient data obtained by modalities such as PET and MRI to develop digital models and phantoms that can be used to simulate the human body function. These are used to obtain the necessary individual equations required to calculate the doses received by each organ from a particular radiopharmaceutical activity administered.

Acknowledgements The material used for this presentation, have been taken from the following sources:

- The SENTINEL European Project, www.dimond3.org
- IAEA, "Nuclear Medicine Resource Manual", STI/PUB/1198, IAEA, Austria, 2006
- The IAEA Nuclear Medicine Educational and Training material available at (http://rpop.iaea.org/RPoP/RPoP/Content/index.htm)

- The EANM web site (www.eanm.org)
- The ESMI web site (www.e-smi.eu)

These are freely accessible and you are encouraged to study and use them.

References

1. Christy M. and Eckerman K.F. (1987) Specific absorbed fractions of energy at various ages from internal photon sources: I. Methods, ORNL/TM-8381/V1, Oak Ridge National Laboratory, Oak Ridge TN.
2. ICRU (1992), International Commission on Radiation Units and Measurements, Phantom and Computational Models in Therapy, Diagnosis and Protection, ICRU Report 48 (Bethesda, MD: ICRU).
3. Kramer, R., et al. (1982), The Calculation of Dose from External Photon Exposures Using Reference Human Phantoms and Monte Carlo Methods, Part 1: The Male (Adam) and Female (Eva) Adult Mathematical Phantoms, GSF-Bericht S-885.
4. Lee, C., et al. (21 September 2006), Whole body voxel phantom of paediatric patients – UF Series B, Phys. Med. Biol. **51**(18): 4649–4661.
5. Schlattl H. et al. (2007), Organ dose conversion coefficients for voxel models of the reference male and female from idealized photon exposures, Phys. Med. Biol. 52: 2123–2145.

Radiation Protection Optimisation in Nuclear Medicine: Therapeutic Procedures

Jim Malone and Geraldine O'Reilly

Abstract Radionuclide therapy has entered a phase in which there is growing interest, with new radionuclides, new labelling techniques and new molecular agents to assist targeting disease sites. This, among other developments, has given rise to questioning of some practices that have become enshrined in radionuclide therapy. In particular the issues around release of the patient back to their community and family has been the subject of much debate. This area will be reviewed and some aspects of unsealed source therapy which may be conducted in a general hospital will be described to illustrate the problems involved and some approaches to their solution.

Keywords Radionuclide therapy • 131-I therapy • Patient release from hospital • Facility design for radionuclide therapy

1 General

Radionuclide therapy, after a period of stability, has entered a phase in which there is growing interest, with new radionuclides, new labelling techniques and new molecular agents to assist targeting disease sites. This and other concerns have given rise to a questioning of some practices that have become enshrined in radionuclide therapy. In particular the issues around release of the patient from hospital, back to their community and family, have been the subject of much discussion and debate.

The main radionuclides presently used for therapy are listed in Table 1. The extent of the radiation protection literature available to support them is highly variable.

J. Malone (✉) and G. O'Reilly
Department of Medical Physics and Bioengineering, Trinity College/St James's Hospital, Dublin 8, Ireland
e-mail: jifmal@gmail.com; jfmalone@tcd.ie

Y. Lemoigne and A. Caner (eds.), *Radiation Protection in Medical Physics*, NATO Science for Peace and Security Series B: Physics and Biophysics 1,
DOI 10.1007/978-94-007-0247-9_12,

Table 1 Radionuclides commonly used for therapy and main emissions

Radionuclide	Main emissions (keV)		Half life (days)
Phosphorus-32	1,710		14.3
Strontium-89	1,492		50.5
Yttrium-90	2,284		2.67
Iodine-131	606	364	8.04
Samarium-153	881	103	1.93
Holmium-166	1,850	81	1.13
Erbium-169	340		9.3
Lutetium-177	500	113, 208	6.7
Rhenium-186	1,070	137	3.8
Rhenium-188	2,120	155	0.7
Gold-198	1,372	411	2.696

Based on IAEA 2009, see reference

Much information is available for the most commonly used radionuclide ^{131}I, which is used both in therapy of hyperthyroidism and thyroid cancer. The literature for many of the others is highly derivative from and dependent on the radioiodine literature. It has recently been reviewed by ICRP, the IAEA and the NCRP.

In this area there are surprising differences in the recommendations available internationally. Those most suited to Europe are those available from the EC, while additional more recent information, compatible with the European approach, is provided in the IAEA publication. The latter seeks to be able to accommodate itself to the statutory framework that prevails in differing jurisdictions. The differences in practice arise from a number of sources including, differences in approach to dose limits and dose constraints for members of the public and comforters/carers. In addition they may arise from the extent to which the patient is involved in the decisions.

On the other hand the physical facilities required for radioiodine therapy of thyroid cancer are, in Europe, much more demanding. The activities involved are high, frequently 1–4 GBq and even up to 8/9 GBq (in the US) for each individual therapy. In Europe this is undertaken on an in patient basis only, with isolation facilities provided for the patient. Generally most of the administered activity appears in urine in the first few days post therapy. For example, one report indicates that 60%, 20% and 5% of the activity appear in urine in the first, second and third days post therapy respectively. Significant activity is also released in sweat and saliva. All of this gives rise to a high potential for contamination.

In addition the external dose rate from the patient could breech dose limits and/or dose constraints for those in their immediate vicinity. This is generally managed by limiting access to the patient and arranging for them to be as self managing as is practical while their retained activity is still high.

Special facilities for unsealed source therapy in radiotherapy departments and general university hospitals have been designed and built in many parts of the world. Those who design and operate these facilities feel they have a satisfactory approach to the radiation protection and clinical problems that arise.

2 Release of Patients Following Radionuclide Therapy

Release of patients after radionuclide therapy had become a black art, using secondary or derived limits that had gradually become distanced from the formal requirements of the regulatory system. In particular ICRP has been concerned that release criteria based on retained activity may be only loosely related to primary dose limits, such as the 1 mSv per year of members of the public. Concern has also been expressed that the methods used to estimated residual activity may on occasions be unduly conservative, and thereby result in unnecessary hospitalization.

The IAEA has devised an approach that is more closely related to primary regulatory requirements and that also takes account of the patient's wishes and his/her family/social circumstances (Table 2). The approach is tailored to each patient and more in keeping with the respect for the individual required in the practice of modern medicine. In addition the conditions that must prevail for involvement of comforters and carers are discussed and appropriate dose constraints are suggested along lines originally proposed by the EU. How to give effect to this approach, including the necessary advice which must be given to the patient and his/her family, is fully described in the IAEA publication. Relevant additional advice and the data to deal with most contingencies that might arise are also provided. This includes advice on how to deal with post mortem procedures, which are increasingly likely to arise, as some patients receiving radionuclide therapy for malignant conditions may be seriously ill at the time of therapy.

Table 2 Issues to be considered when releasing patients

Patient and institutional issues to be considered in release decisions
Patient issues in release decision
Their medical needs
Their wishes
Their pattern of contact with other people
Their age
Their family/home environment
Aspects of lifestyle involving occupational/public exposures
Cost/environmental factors
Their local social and infrastructural arrangements
Issues to be considered in release decision
Requirement for regulatory compliance, based on
Isolation of the patient to reduce dose to the public and family
Issues associated with the patient's medical condition that require hospitalization, or might reduce compliance
A requirement to collect and store of urine to reduce radioactive discharges into the sewer and/or the impact of incontinence

Adapted from IAEA Publication cited below

References

1. ICRP (2004). International Commission on Radiological Protection, Release of Patients After Radionuclide Therapy, Publication 94. Oxford, Elsevier.
2. NCRP (2006). National Council on Radiation Protection and measurements. Report 155: Management of radionuclide therapy patients. Bethesda, MD, NCRP.
3. EC (1998). European Commission, Radiation Protection 97, Radiation Protection Following Iodine-131: Luxembourg, European Commission.
4. RCP (2007). Royal College of Physicians, Radioiodine in the management of benign thyroid disease, Clinical guidelines, Report of a Working Party. London, RCP.

Radiation Protection of Pregnant Women

Jim Malone and Geraldine O'Reilly

Abstract One of the more sensitive problems in radiation protection arises when the person irradiated is, or is possibly pregnant. This paper deals with the risks to the foetus and the regulatory environment that arises when a pregnant person is irradiated.

Keywords Pregnancy • Pregnant women • Radiation • Patient • Consent • Information

1 Introduction

One of the more sensitive problems in radiation protection arises when the person irradiated is, or possibly is pregnant. This paper deals with the risks that may arise for the foetus when a pregnant person is irradiated. It considers these risks in the context of both the radiobiology and the normal progress of pregnancy. In addition it considers the regulatory environment for the different groups involved and the advice that may be offered.

2 Radiobiological Considerations

During a normal pregnancy there is a real risk that the pregnancy will fail and/or a spontaneous abortion may occur. The risk of this is generally believed to be greater than 15%. In addition there is risk of genetic abnormalities, growth retardation, and the

J. Malone (✉) and G. O'Reilly
Department of Medical Physics and Bioengineering, Trinity College/St James's Hospital, Dublin 8, Ireland
e-mail: jifmal@gmail.com; jfmalone@tcd.ie

Y. Lemoigne and A. Caner (eds.), *Radiation Protection in Medical Physics*,
NATO Science for Peace and Security Series B: Physics and Biophysics 1,
DOI 10.1007/978-94-007-0247-9_13,

Table 1 Risks to the Embryo

Risks
Pre-natal or neo-natal death
Congenital abnormalities
Growth impairment
Reduced intelligence
Genetic aberrations
Increase in risk of cancer

Table 2 Pattern of incidence for deterministic effects

Weeks of gestation	Deterministic effects
0–3	Failure to implant/death of embryo
3–8	Organ malformation
8–25	IQ Shift, 30 points per Sv weeks 8–15; smaller in next 10 weeks
>25	Lower risk period

possibility of major malformations as the pregnancy progresses. The natural incidence of the latter is believed to be in the range 2–4%, depending on the definition used.

Risks to the embryo, from early irradiation, are listed in Table 1. As pregnancy proceeds, some of these risks decline. The pattern of incidence for deterministic effects is summarized in Table 2. The lower value for the threshold for malformations is 100 mGy. This value is also frequently used for IQ shifts although higher values, up to 300 mGy, are occasionally quoted.

In addition stochastic effects, mainly leukaemia and cancer, arise from irradiation of the foetus, at a rate estimated to be roughly equivalent to that which prevails with the irradiation of young children. The HPA quotes an increased excess absolute childhood cancer risk coefficient of 1 in 13,000 per mGy. To place this in context, the natural incidence of cancers up to age 15, in the UK, is 1 in 500. In terms of fatal cancers, 25 mGy will double the natural risk. ICRP takes the view that there are no known effects in human populations from preconception irradiation of either parent at the doses involved in diagnostic procedures.

With all of the above, irradiation of the foetus adds to the natural risks. At low doses it will be impossible, in practice, to distinguish a case in which one of the phenomena listed has occurred naturally from a case in which it has been radiation induced. This creates a need for skilled communication with the mother, by a person she can trust who can combine empathy, insight and scientific knowledge.

3 Regulatory and Good Practice Considerations

The regulatory framework for irradiation of pregnant and potentially pregnant patients (and carers) has been set out in the Medical Exposures Directive (MED), with additional information provided in EC Publication RP 100. When the person irradiated is a member of staff the EU BSS applies.

With regard to dose limits, the most important additional consideration is that the dose limit generally applied to the foetus is that which prevails for the general public. In Europe this is 1 mSv after the declaration of pregnancy, although higher values are used elsewhere. Thus pregnant workers, in most countries, must follow a regime that ensures the dose limit for workers is met and, in addition, the more demanding dose limit for the foetus is met. This may require special provision for monitoring pregnant workers. For workers in medicine, it is normally possible to ensure these dose limits are met.

In addition, other legislation dealing with the general arrangements for pregnant workers has been enacted in many countries. This may have a significant bearing on how the radiation related provisions are applied. For example, some countries require the employer to "assess risk to safety.... and take the necessary preventative and protective measures". In addition they may require that "If ... preventative measures can not remove risk then the working conditions or working hours should be temporarily adjusted" and finally they can require that "If adjustment of working conditions or hours is not possible then employees should be provided with other work." However, the latter is generally not necessary with medical uses of radiation, except in unusual and exceptional circumstances.

With regard to irradiation of patients, there is great variation in practice throughout the world and throughout Europe as demonstrated by recent surveys. Some of the variation is at least questionable, and may not be consistent with the published advice. This has become a matter for concern as the number and frequency of examinations that deliver high doses to the foetus increases. These include examinations in which the foetus can be partially or completely included in the examination field, such as abdominal/pelvic x-rays, barium meals and enemas, and CT examinations of the abdomen/pelvis. Foetal doses in the range 1 – 80 mGy are seen with these examinations, and are a cause for concern. When the foetus is irradiated during pregnancy, a dose and risk assessment should be made, appropriately documented, and communicated to the relevant parties. Where the exposure(s) are deliberate they must be planned to achieve the diagnostic task and to minimise foetal exposure.

The measures taken to reduce unnecessary foetal irradiation include protocols to reduce or eliminate examinations that deliver significant doses to pregnant or potentially pregnant individuals, except when the risk of foetal irradiation has been explicitly justified by the prescriber or the practitioner. In addition it is highly recommended that, where possible, the mother/potential mother be closely involved in the decision and informed consent be obtained. Some countries use protocols such as the 10-day or 28-day rules to assist with this, others use pregnancy tests, while others take no action. Some of the more commonly used and easily available pregnancy tests can be misleading and, with negative results, give a false sense of security. The reasons advanced for the variability in practice and for not taking action and/or not seeking the woman's consent are unlikely, should they be challenged, to be accepted by the public.

Some patients feel a termination may be required if they undergo a diagnostic radiology procedure during pregnancy. In many of these cases lack of knowledge is responsible for great anxiety and contributes to unnecessary terminations. For most

Table 3 Additional advice

Foetal dose and terminations
• Termination at fetal doses of less than 100 mGy is not justified based upon radiation risk
• At fetal doses in excess of 500 mGy, there can be significant fetal damage, the magnitude
• Type of which is a function of dose and stage of pregnancy
• At fetal doses between 100 and 500 mGy, decisions should be based upon individual circumstances, and credible, experienced counselling which patient will find trustworthy should be provided

patients foetal risk is minimal. It is only in exceptional circumstances that it might be necessary to consider termination following diagnostic exposure. In Table 3 additional advice is provided for situations where the foetal dose may be higher, which is possible with interventional procedures and multiple CT scans. For reports which develop these points more fully, see the references below.

References

1. HPA (2009). Protection of Pregnant Patients during Diagnostic Medical Exposures to Ionising Radiation. *Doc HPA* RCE-9. Available from http://www.hpa.org.uk/webw/HPAweb&HPAweb Standard/HPAweb_C/1238230848780?p=1199451989432.
2. ICRP (2001). International Commission on Radiological Protection. Pregnancy and Medical Radiation, Pub. 84, Annals of the ICRP 30(1).
3. EC (1998). EUROPEAN COMMISSION, Radiation Protection, RP 100: Guidance for protection of unborn children and infants irradiated due to parental medical exposures, Office for Official Publications of the European Communities, Luxembourg.
4. Schreiner-Karoussou A (2008). Review of existing issues and practices with respect to irradiation of patients and staff during pregnancy. Radiat Prot Dosimetry 129(1–3): 299–302; (2009) A preliminary study of issues and practices concerning pregnancy and ionising radiation. Radiat Prot Dosimetry:135(2): 79–82.

Optimization of Dose Distribution

Pawel Kukolowicz

1 Introduction

Radiotherapy, if used appropriately, constitutes a very effective treatment for cancer. It is considered as the second, after surgery, most effective type of cancer treatment. In developed countries more than half of all cancer patients are treated with ionizing radiation. Radiation therapy is used both for radical and palliative treatment. The radiation affects only the tissues in the treated area and not the rest of the body. Therefore, it is known as a local method of treatment. Radiotherapy, like any other form of treatment, has risks and benefits. The risks are associated with the dose delivered to normal organs and tissues. Therefore, every application of radiation for the treatment of patients suffering from cancer has to be optimized. The goal of radiotherapy is to deposit the prescribed dose in a cancer tumour while minimizing the dose that reaches the surrounding normal tissue. To achieve this aim, the spatial dose distribution should conform to the target volume, and the gradient of the dose distribution outside of the target should be as high as possible – the dose should quickly falls off.

Figure 1 shows the dependence of the tumour control probability and the normal tissue complication probability for a typical situation occurring in radiotherapy.

In the case of the clinical situation presented in Fig. 1 in order to cure 100% of the patients a dose of about 100 Gy has to be delivered to the tumour. Unfortunately, at this dose, serious adverse effects will be observed in almost each patient. If we accept that in 5% of patients the normal tissue damage may appear when a dose of about 62 Gy is delivered to the target, then this dose allows tumour control in more than 60% of the patients. To keep the NTCP at the level of 5% and to increase the probability of tumour control (in most clinical cases the 5% is the acceptable risk of tissue complications) a better dose distribution is required. The aim is, as it has already been mentioned, to deliver the prescribed dose to target while minimizing

P. Kukolowicz (✉)
Medical Physics Department, Holycross Cancer Center, Kielce, Poland
e-mail: pk@message.pl

Y. Lemoigne and A. Caner (eds.), *Radiation Protection in Medical Physics*,
NATO Science for Peace and Security Series B: Physics and Biophysics 1,
DOI 10.1007/978-94-007-0247-9_14,

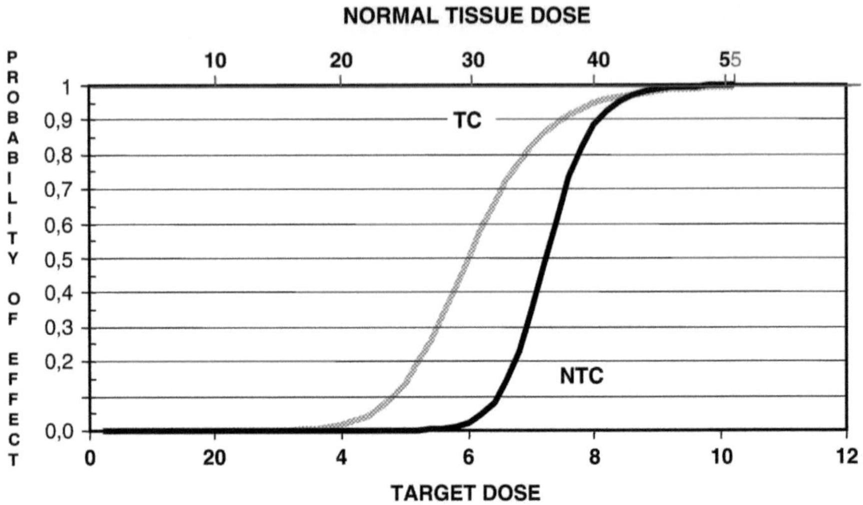

Fig. 1 TCP and NTCP – dependence on dose

the dose to normal structures. Technically, we may say that optimization of dose distribution relies on moving the curves of TPC and NTCP apart from each other as far away as possible. There are several methods that help to obtain better dose distribution. Some of them will be presented below.

2 Dose Distribution

Dose distribution is usually presented in the form of isodoses. The isodoses are the lines composed of points in an absorber that receive equal doses of radiation. It is generally accepted that the dose distribution should fulfil at least two conditions: (1) the dose distribution in the target should be as homogenous as possible, (2) the gradient of dose distribution outside of the target should be as high as possible. The aim of radiotherapy is to kill all malignant cells. The higher the dose the larger the chance to reach this aim, and the more malignant cells present in the target the higher dose should be delivered. There is no precise information on the distribution of malignant cells in the body. From pathology it is known that in different part of the body there is a different number of malignant cells. Then it is reasonable to match the delivered dose to the expected number of malignant cells. A larger dose to where there are more malignant cells and a smaller dose to where there are less malignant cells. Saying that the dose distribution should be as homogenous as possible means that in each sub-volume the dose distribution is homogenous. Figure 2 shows such clinical situation where three different sub-volumes with different number of malignant cells are indicated. In each sub-volume the dose distribution should be homogenous but the total dose delivered could be different.

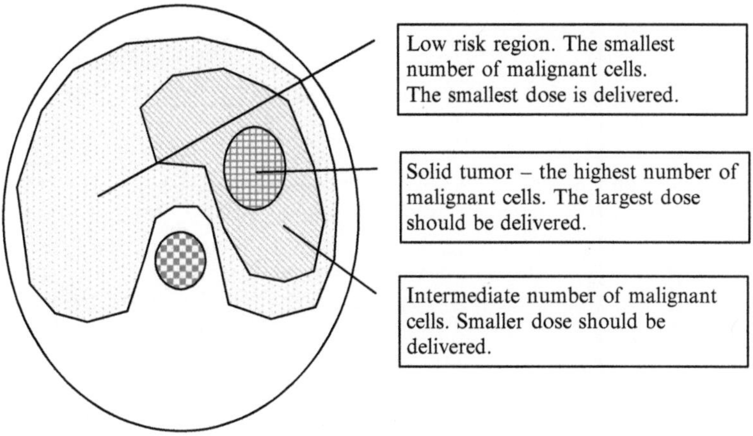

Fig. 2 The low, intermediate and high risk regions of target volume. In each sub-volume dose distribution should be homogenous, albeit different doses will be prescribed to these regions

There are two commonly used metrics of uniformity of dose distribution in the target volume. The first one is proposed by the International Commission on Radiation Units and Measurements (ICRU). Let us represent the dose delivered to the target in percentages of dose prescribed by a radiotherapist. The dose distribution is acceptable, if doses in the target are in the range 95–107%. The second one is proposed by the Nordic Association of Clinical Physics (NACP). The uniformity of dose distribution in the target is given in the form of standard deviation of dose distribution. The dose distribution is acceptable, if the standard deviation is smaller than 3%.

3 Choosing the Appropriate Radiation and Its Energy

The dose distribution depends on the type of radiation applied and its energy. Two different types of radiation are used most often in radiotherapy. These are high energy X-Rays and electrons. In some countries also protons and very high energy ions are used. The application of these particles is limited due to their large cost. Here, we will focus on photons and electrons. Figure 3a and b show the dose distribution of photons and electrons in water as a function of depth (dose distributions are measured in water because interaction with a treatment beams with water is very similar to interaction with human soft tissue) respectively.

The range of photons is much higher than electrons so photons are well suited for irradiation of deeply located tumours while electrons are good for superficial lesions. The second very important feature of photons is a typical build-up region close to entrance to the absorber. When the beam enters the water (the human body) the dose increases very quickly with depth reaching the maximum at a depth, which

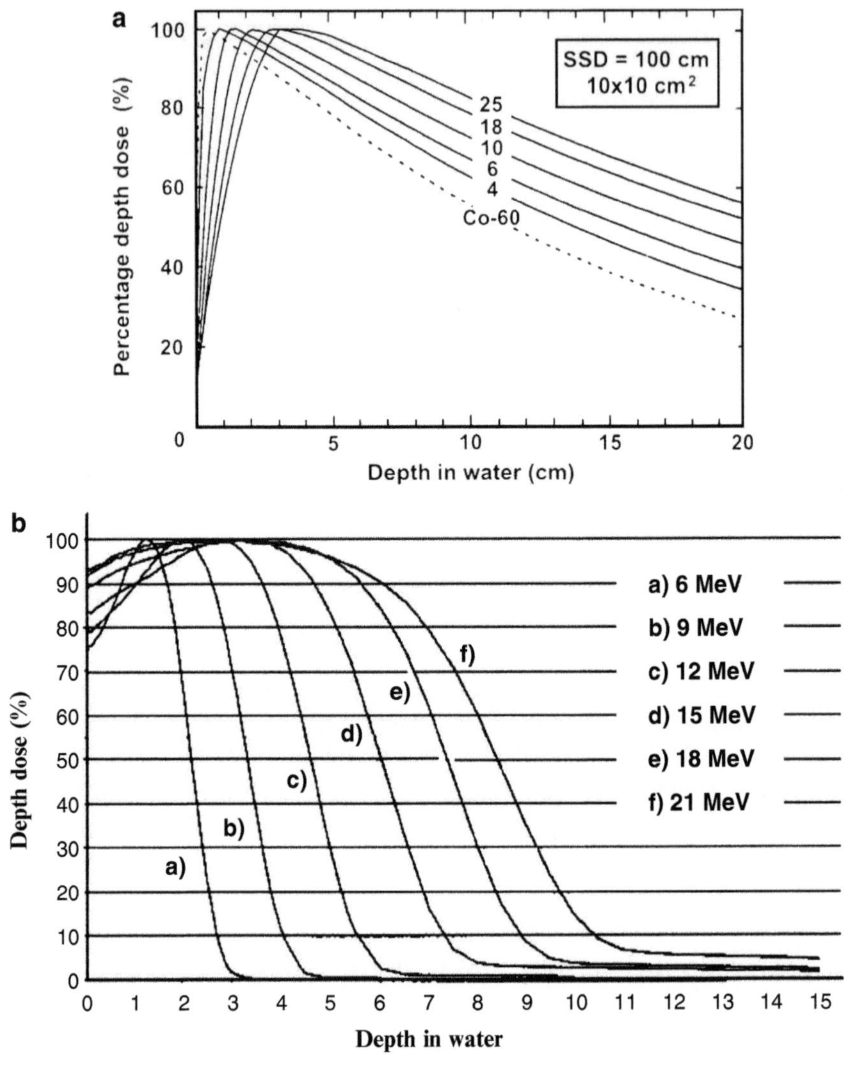

Fig. 3 (**a**) Depth doses for photons – Co60 beam, 4, 6, 10, 18 and 25 MV X-Rays. The depth doses were measured for a square field of 10 cm. Dose distributions were measured at 100 cm distance from the source to the surface of the phantom. The distributions are normalized to maximum dose (to dose at a point where the maximum dose is delivered). (**b**) Depth doses for electrons of energy 6, 9, 12, 15, 18 and 21 MeV. The depth doses were measured for a square field of 10 cm. Dose distributions were measured at 100 cm distance from the source to the surface of the phantom. The distributions are normalized to maximum dose

depends on the energy of the beam. The larger the energy the deeper the maximum dose is positioned. Build-up of high energy photons allows avoiding serious injuries of skin. For energies up to 400 kV X-rays (orthovoltage radiation) the maximum dose was delivered to skin, which resulted in injuries of skin and even necrosis.

The deeper the tumour is in the body, the higher energy of photons should be used. However, if more photon fields are used, then smaller differences are observed in the dose distribution obtained. The largest difference is seen for the so called two opposed fields technique.

Figure 4 shows the dose distribution along the central axis for gamma radiation of Co60 1.25 MeV, and for two linear accelerators of energies 6 and 15 MV.

As may be noticed, the smaller the energy, the greater dose is delivered at points lying farther from the centre of the target. For higher energies the dose distribution is quite homogenous except for the regions very close to the surface, where the dose is much smaller. For Co60 the dose at a depth of 1 cm the dose is of about 16% higher than the dose at the mid point.

Electron beams are used for relatively superficially located tumours. In most cases, if an electron beam is used, then a single beam is used. It is widely accepted that the dose to the tumour cannot be smaller than 80% of the maximum dose. For electrons the depth at which the dose decreases to 80% is sometimes called the therapeutic range. The therapeutic range is given by the formulae:

$$R_{80\%}[cm]=\frac{E[MeV]}{3}$$

For example, tumours which are located not deeper than 4 and 5 cm respectively beneath the skin surface may be treated with 12 and 15 MeV electron beams, respectively. No matter how useful the electrons seem for treatment of shallowly located tumours two facts should be taken into account. The first is a relatively high dose delivered to skin which may lead to a serious injury of this tissue, as in the case of the orthovoltage radiation. The dose to surface is for all electron energies larger than 80% of dose to maximum. The second one is that electrons are very

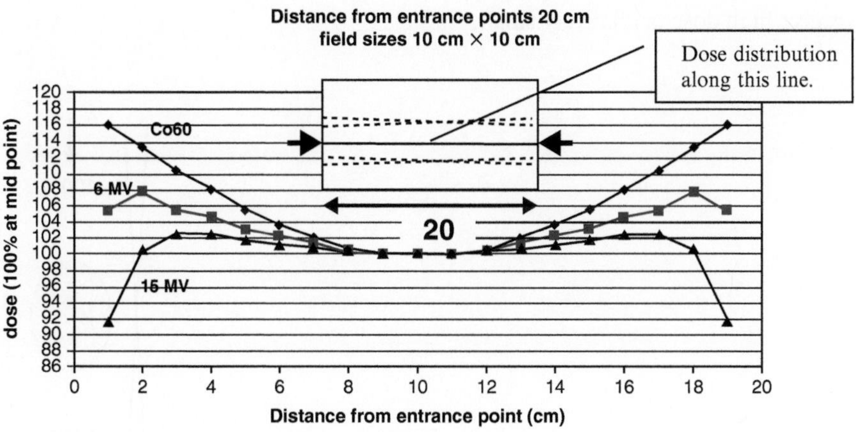

Fig. 4 Dose distribution for two opposed beams. The small thick arrows indicate the place of beam entrance points

easily scattered while they are travelling through the tissue, especially when they pass through tissues of different densities. This may lead to a quite inhomogeneous dose distribution. Therefore, for electrons it is rather difficult to obtain a dose distribution which fulfils the requirements of ICRU and NACP.

4 Choosing the Appropriate Number of Beams and Beam Orientations

The selection of a set of suitable beam orientations is the most powerful tool in dose distribution optimization. In conventional treatment planning several trial-and-error attempts are used to determine a set of adequate orientations. An experienced physicist or anybody involved in treatment planning after analysis of the target position with respect to organs at risk, is usually able to decide how many fields should be used and which orientations are the best. There are also some proposals of how to automate the selection of beam orientations. A major obstacle in applying these methods is their excessive computation time. Here we shall describe some simple rules which may help in optimization of dose distribution by appropriate selection of beam orientations.

4.1 Number of Beams

Figures 5–8 show the approximate dose distribution for several beam angles.

The dose outside the target is very close to the dose delivered to the target. All organs at risk placed inside of the fields absorb a dose very close to the prescribed one. This situation was previously analyzed in Fig. 3. The volume of tissues which receive high dose is relatively small.

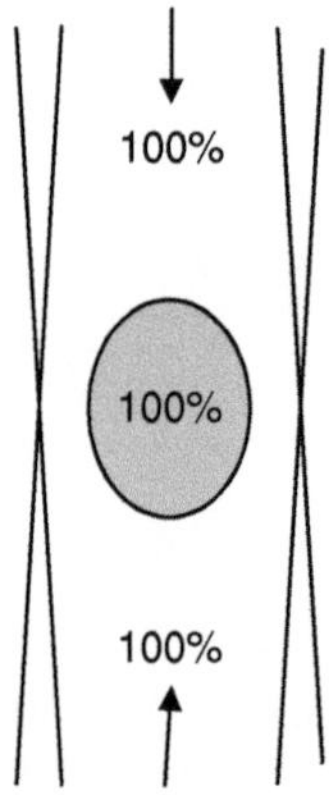

Fig. 5 Dose distribution for two opposed beams

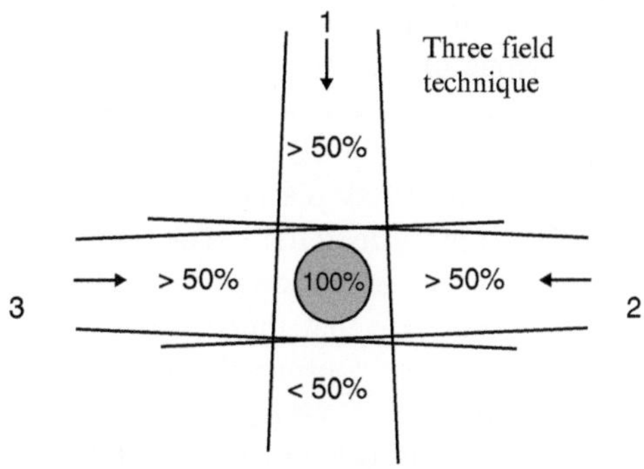

Fig. 6 Dose distribution for two opposed beams

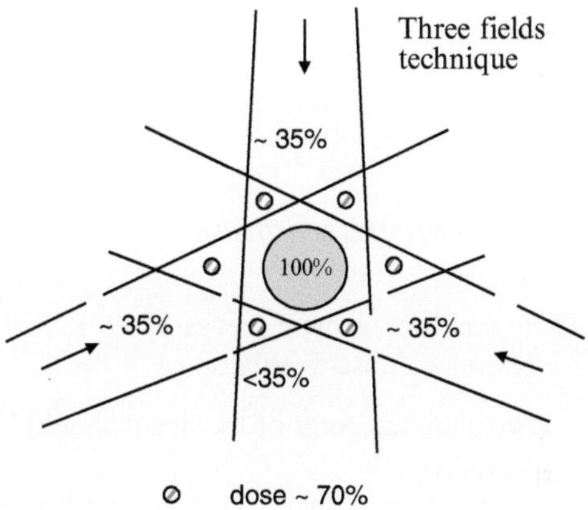

Fig. 7 Dose distribution for three field technique. Dose distribution for three field techniques – beams at 0, 120 and 240 angles

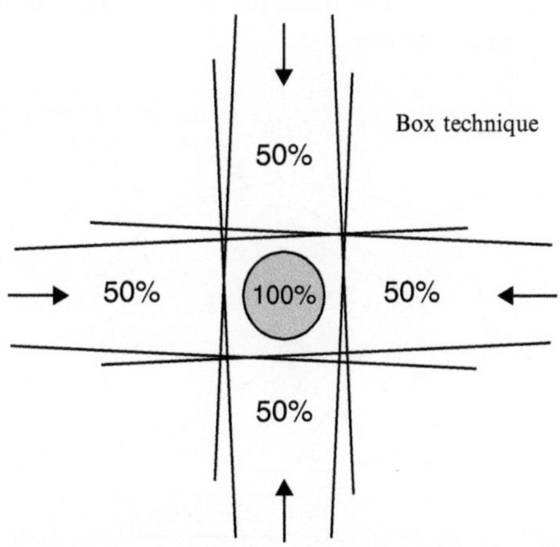

Fig. 8 Two pairs of two opposed beams. The so called box technique

Dose distribution for three fields techniques – one horizontal and two lateral fields. None of the organ at risk placed outside the target receives a dose much larger than 50% of the prescribed dose. The volume of tissues which receive high dose is larger than in the case of two opposed beams.

All organs at risk placed in the fields and outside of the target absorb about 50% of the prescribed dose. The region of high dose increases in comparison to the two opposed fields and three fields technique. From this very simple analysis we may conclude that the increase of the number of beams makes the dose distribution more conformal – the gradient of dose distribution outside of the target is higher.

The price paid is the increase of the region of relatively high dose, e.g. for box technique this region is approximately two times larger than for two opposed beams. Making the right decision is always more complicated and it depends on the actual situation. However the analysis presented here might be helpful. To be an expert in treatment planning, the dose distributions for these simple techniques must be known by hard.

So far the importance of the number of beams has been analyzed. Some general suggestions concerning the beam orientation may be given.

4.2 Beams Orientation

There are three general pieces of advices, which if followed, may help in getting the optimized dose distribution:

1. The entrance point of the beam should be as close as possible to the centre of the target
2. The direction should be chosen in such a way that the beam will omit organs at risk
3. The direction should be chosen in such a way that the cross section of the beam will be as small as possible

To explain these three pieces of advice let us analyze a clinical situation presented in figures below.

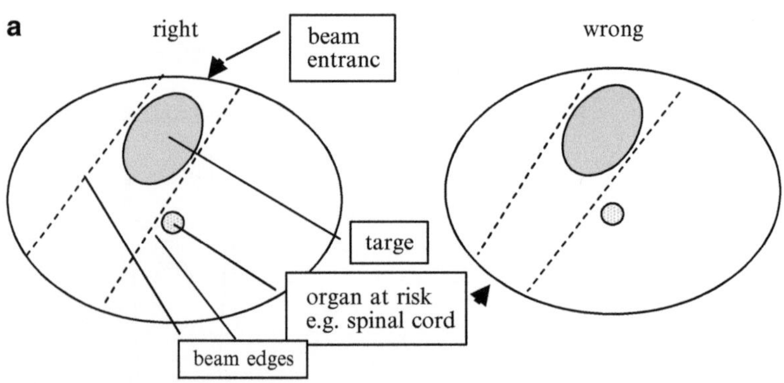

b

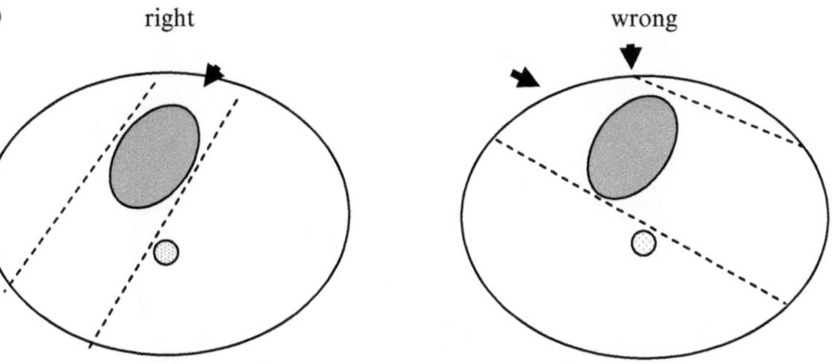

c

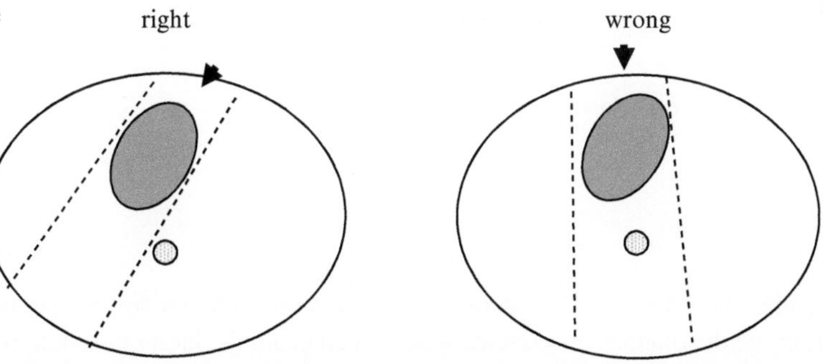

These proposals are quite often contradictory, so that an individual decision must be made.

5 Uniformity of Dose Distributions in the Target Volume

It has already been mentioned that in classical radiotherapy a uniform dose distribution is an important task in the optimization of dose distribution. Figure 9 shows schematically the three field technique. Vectors in the centre of target (point Q), denoted as $\nabla D1$, $\nabla D2$and $\nabla D3$, describe the gradient of each single beam's dose distribution. It was assumed that from each beam 1 Gy is delivered to the point Q. Let us assume that the prescribed dose is D_{pr}.

In the first approximation to receive a homogenous dose distribution the following relation must be fulfilled:

$$w1 \cdot \nabla D1 + w2 \cdot \nabla D2 + w3 \cdot \nabla D3 \cong 0$$

where w1, w2 and w3 are the weights of the beams. To deliver the prescribed dose, the following relation must be fulfilled:

$$w1 + w2 + w3 = D_{pr}$$

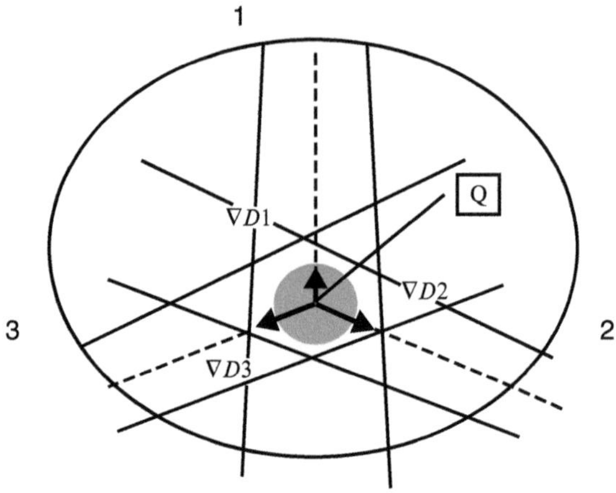

Fig. 9 How to make the dose distribution homogenous

The reader should notice that there is more than one solution for this set of linear equations so the planner may decide from which beam the larger dose will be delivered. In this way one may decrease the dose delivered to an organ at risk which cannot be fully omitted.

Modifiers called wedges are used for improving the uniformity of dose distribution. They are especially well suited for two opposed fields technique. With wedge, the dose distribution for a single beam is modified. Figure 10 shows the dose distribution for open field (without wedge) and with wedge of angle α. For an open field the isodoses are perpendicular to the central axis in the central part of the beam. With the application of a wedge, the angle between the isodoses and central axis may be changed.

6 Summary

The optimization of the dose distribution allows for the safe and effective treatment with ionizing radiation. There are several tools which may be used in the optimization process. The most important are the right choice of the number of beams and

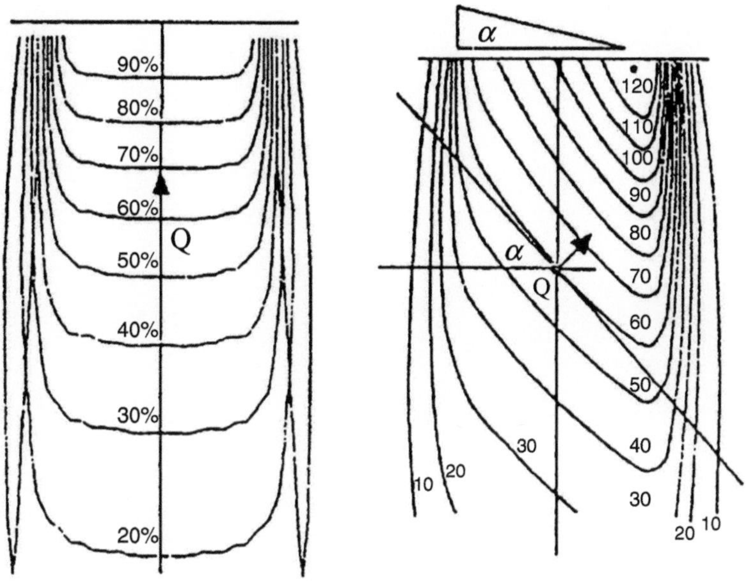

Fig. 10 Dose distribution for open and wedge fields

their orientations. There is also a very sophisticated method of dose distribution optimization, namely the so called Intensity Modulation Radiotherapy. This method has not been presented in this text, partly because in the author's opinion, the conventional methods, if applied appropriately are very powerful. The knowledge of the dose distribution for a single beam of a given radiation is a prerequisite for dose distribution optimization.

Optimisation of Radiation Protection in Diagnostic Radiology

Cornelius Lewis

1 Introduction

The basic framework for Radiation Protection, according to ICRP, is based on three fundamental principles; justification, optimisation and limitation. The second principle, optimisation, is where the medical physicist can have the greatest influence. Briefly, the principle states that doses should be kept *As Low as Reasonably Achievable (ALARA)*.

Why is ALARA important? The effects of radiation on living organisms lead to stochastic or deterministic effects. The stochastic effect, associated with genetic mutations, may lead to cancer induction and the evidence indicates that stochastic effects follow a linear no-threshold (LNT) model. The most recent data supports this model down to dose levels as low as 20 mSv (2 rad) which is within, although at the upper end, of the dose range experienced in diagnostic radiology [1].

2 Routes to Optimisation

There are a number of approaches to optimisation which may be sub-divided into two general categories; design and function of equipment and imaging techniques.

Optimisation through design and function is influenced by the choice of equipment and materials and the subsequent quality assurance and control measures that are applied once the system is operational.

The choice of imaging technique and the imaging protocols selected are the areas which will have the greatest influence on patient dose and should be an essential element of optimisation activities.

C. Lewis (✉)
King's College Hospital, London, UK
e-mail: cornelius.lewis@kcl.ac.uk

Y. Lemoigne and A. Caner (eds.), *Radiation Protection in Medical Physics*,
NATO Science for Peace and Security Series B: Physics and Biophysics 1,
DOI 10.1007/978-94-007-0247-9_15,

3 Specification, Acceptance and Commissioning

When new equipment is selected the focus is on capability and 'fitness for purpose'. However, even at this stage, choices may be available to ensure that equipment is both fit for purpose and able to operate in a dose efficient manner .

An obvious example of this is the choice of materials used in construction. X-ray techniques rely on attention within the patient but the x-ray beam must also pass through other structures before the image is captured. The simplest example is the x-ray table. Modern systems are designed with carbon fibre tables. This material is very strong yet it does not add significant attenuation to the x-ray beam. Similarly, x-ray grids and cassettes are also available with carbon fibre components.

Filtration plays an important role in reducing patient dose. All systems will be fitted with aluminium filtration to reduce the low energy component of the x-ray beam which has insufficient penetration to play a useful part in image formation. The total filtration (inherent plus added) of an x-ray system is set at a minimum of 2.5 mm aluminium. However, there is increasing interest in the use of additional filters, usually copper, which may be employed in certain procedures to reduce unnecessary dose [2, 3].

The use of film to record x-ray images is becoming redundant and being replaced by digital image capture systems, either Computed Radiography or Digital Radiography. Digital image capture systems will differ from film in their response to radiation and if this is not taken into consideration when such systems are used the opportunity for dose reduction may not be realised.

4 Quality Assurance and Quality Control

X-ray systems must receive regular performance checks to ensure that they are operating optimally. Parameters that should be checked include; filtration, kV, mA, time and, in addition for fluoroscopy systems, contrast, resolution, image intensifier dose rates etc. Computed tomography procedures deliver some of the highest patient doses in diagnostic radiology and these systems should also be subject to regular quality checks.

The UK Institute for Physics and Engineering in Medicine has published guidance on quality assurance checks (IPEM 91) [4]. This guidance specifies the performance parameters that need to be checked, the expertise required to perform the measurements, the frequency with which the measurements should be undertaken, the relative importance (priority) of the parameter and, importantly, tolerance levels beyond which action should be taken.

Tolerance levels are usually set at ±10% for remedial action and ±20% for suspension. In both cases it would be necessary to contact the equipment service engineer to arrange for a repair. However if the tolerance exceeds the suspension level it should be recommended that the equipment be removed from operation until any fault has been rectified.

Quality assurance and control should be a shared responsibility between radiographers and medical physicists. The IPEM 91 scheme refers to Level A and Level B expertise. The intention is that simple, Level A, checks should be undertaken frequently to provide reassurance that the equipment is operating to specification. Should a problem be identified following simple quality checks it would be necessary to contact medical physicists to undertake more detailed checks. Medical Physicists should also undertake more detailed equipment quality checks (Level B) on a regular but less frequent basis.

A particular example of the type of measurements that could be made at Level A are those with a simple exposure meter. If such measurements are made using a reproducible set up any variation in kV, mA, time, filtration etc will lead to a different exposure reading. Measurements can easily be taken on a daily or weekly basis and tabulated or plotted (see next page). If the recorded measurements fall outside a given range, for example ±10%, action should be taken. This action would normally be to inform the medical physics service or service engineer (Fig. 1).

A key device for ensuring optimisation in x-ray systems is the Automatic Exposure Control (AEC). In its simplest form an AEC is a parallel plate ionisation chamber placed immediately in front of the image receptor. When sufficient charge is collected by the AEC, a signal is sent to the operating console to terminate the exposure. In most systems the AEC contains three chambers which may be selected individually or in groups.

The level of charge which causes the exposure to terminate is normally set and adjusted by the service engineer although minor adjustments are available to the user. The AEC also has a 'back-up' timer associated with it so that in the event of a failure of the AEC chamber to terminate the exposure, it will automatically terminate once the back-up time is reached.

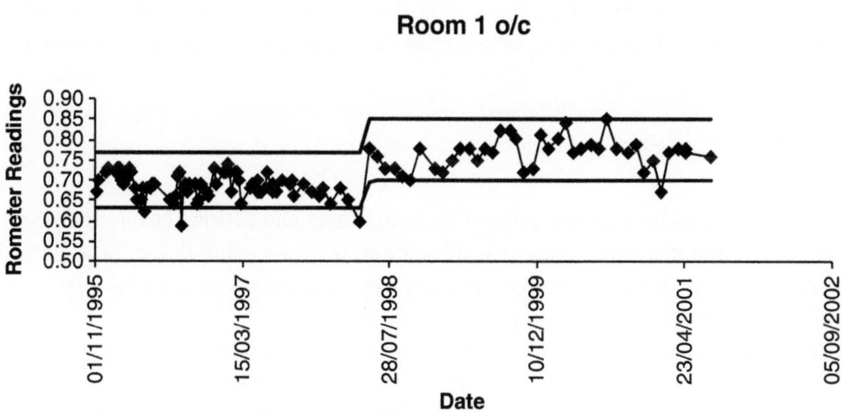

Fig. 1 Example of routine Level A measurements

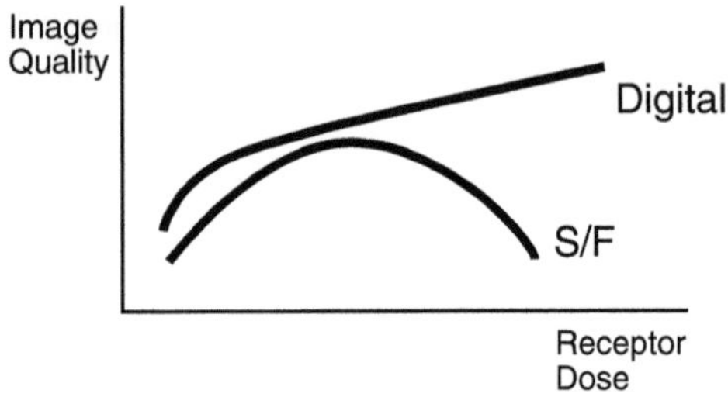

Fig. 2 Image quality versus receptor dose for digital and film radiography

It is essential for the AEC to be tested routinely by medical physics staff – a level B test. There are effectively three tests that need to be performed

1. The sensitivity of all chambers in the AEC should be matched.
2. The image contrast obtained from a standard phantom when the AEC is operating should be adequate for diagnosis.
3. The back up timer should operate and should be set at a realistically low time.

As digital systems gradually replace film for plain radiography it becomes even more important to ensure that the AEC works effectively. The reason for this is the non-linear characteristics of the optical density/exposure relationship for film. There is a relatively narrow range of exposures over which diagnostic images can be obtained (see Fig. 2).

In contrast, for digital image receptors, once a threshold dose is exceeded image noise decreases and therefore image quality increases over a wide dose range, far greater than with film (see figure above).

AEC systems are also being introduced into CT scanners. Manufacturers refer to them by proprietary names such as; Smart mA, Caredose, Sure Exposure. Scanner exposures can be varied at three different levels. At the most basic level adjustments can be made for overall patient size. Progressing from this, the next level is to vary along the z-axis according to variations in patient thickness. Finally some systems are capable of modulating exposure as they rotate in an individual slice. These systems provide opportunities for considerable dose saving in CT. The role of the medical physicist with these systems is to ensure they are used effectively if provided.

5 Image Criteria

Quality assurance and quality control measurements and actions ensure that x-ray generating equipment performs to specification. It is an essential and basic part of the optimisation process but, arguably, the way equipment is used has the most effect on patient dose and thus optimisation.

At a most basic level the way that images are viewed can make a difference to diagnostic quality. Optimal viewing conditions require the image to be well illuminated whilst stray illumination – from windows, room lights, light leakage around images etc – should be minimised. Ideally images should be viewed, whether film or VDU, in a darkened room with the viewer suitably dark adapted.

Performance of viewing boxes or screens should form part of the overall optimisation programme. Simple checks with light meters can be performed to ensure that the display unit (whether a light box or VDU) provides a uniform luminance. If a VDU unit is being used, the supplier should provide test patterns which will enable the contrast and brightness to be set to manufacturers specifications. It should also not be forgotten that the electrostatic charge that builds up on a VDU screen will collect dust avidly. (Clean your own computer screen and observe the improvement in image contrast!).

An extremely difficult aspect of diagnostic imaging is deciding the appropriate imaging parameters to obtain the required diagnostic result. There will always be an amount of personal choice in this but standard protocols have been developed for a range of routine procedures. Three documents have been produced by the European Union detailing imaging guidelines for diagnostic radiographic imaging in adults, children and for CT [5–7]. The latter document has recently been superseded by a series of guidelines and protocols which are the result of an EU funded project on Multi Slice CT Scanning (Safety and Efficacy) [8].

The standard diagnostic protocols consider what structures need to be visualised and propose a series of parameters to achieve the desired image quality together with an estimate of the patient entrance surface dose.

One of the main drivers of patient dose in CT scanning is image noise which is inversely proportional to dose. Noise affects resolution and becomes a significant problem with small structures. Conversely, imaging of large structure with low noise protocols will deliver unnecessarily high doses to patients.

6 Audit

Once work has been completed to optimise equipment performance and imaging protocols a programme of routine clinical audit should be established to ensure that optimal conditions remain. The EU Medical Exposures Directive [9] requires all medical establishments within the EU to develop Diagnostic Reference Levels (DRLs). These should be established locally, usually at the 75% percentile of the dose distribution, but must be with reference to national (or international) recommended DRLs.

The usual parameters used to compare Diagnostic Reference Levels are Dose Area Product (DAP) or Entrance Surface Dose (ESD) in plain radiography or fluoroscopy and $CTDI_{vol}$ or DLP (Dose Length Product) in CT. It is becoming common for all equipment to provide displays of these parameters making the process of audit relatively straight forward. However a range of other, easily accessible, parameters may also be used equally well. An obvious and simple example is fluoroscopy time.

Whatever parameter is chosen for audit, data should be sampled on a regular basis (perhaps annually) and compared against the standard. If the results indicate that a the DRL has been exceeded on a regular basis an investigation should be undertaken to establish the reason. It is also possible, perhaps through the introduction of new technology that doses are all below the DRL in which case consideration should be given to reducing it.

7 Training

In addition to the requirement for DRLs to be established, there is a further requirement for staff who are practically involved in exposures to have received suitable training in radiation protection. The EU has published guidelines on the theoretical training required amongst different staff groups [10].

The focus of training is often on theoretical training because it is relatively easy to organise. However a very elegant piece of work published in the British Medical Journal in 2003 indicates that theoretical training is of little residual value [11].

Training should focus on practical issues – operating the equipment, determining imaging protocols etc to be of real value.

8 Economic Factors

The formal definition of optimisation ends with the phrase 'economic and social factors taken into account'. This implies that there are constraints on the amount of optimisation that can or should be reasonably expected. Spending large amounts of money to achieve a very small dose saving is not considered beneficial to society as a whole.

The formal process of determining how much should be spent on dose reduction is Cost Benefit Analysis. This compares the cost of introducing optimising features, eg DAP meters, additional shielding etc, against the dose saving that may be achieved. The measure of dose saving is determined as a 'population dose' expressed in man Sieverts. For example, the introduction of a technology that would save 0.1 mSv for exposures undertaken on one million people each year would be 100 man Sv per annum.

To undertake such analysis requires a cost to be associated with population dose and the generally quoted amount is in the region of €50,000–75,000 per man Sv. In Cost Benefit Analysis theory using the example given above, an expenditure of €5–7.5 million per annum might be justified.

A further and more general development of Cost Benefit Analysis is Cost Utility Analysis which uses a quantity referred to as the QALY – Quality Adjusted Life Years. This more detailed analysis takes account not only of the benefit but how the benefit may be realised in terms of the quality of life following the intervention. Cost Utility Analysis is used far more widely in medicine.

References

1. International Agency for Research on Cancer (IARC) 15 Country Study, Cardis et al. 2003.
2. Nicholson et al., Br.J.Radiol., 73, 2000.
3. Moore et al., Br.J.Radiol., 81, 2008.
4. Recommended Standards for the Routine Performance Testing of Diagnostic X-Ray Imaging Systems, Institute of Physics and Engineering in Medicine (IPEM), 2005.

5. European Guidelines on Quality Criteria for Diagnostic Radiographic Images, EUR 16260, ISBN 92-827-7284-5, 1996.
6. European Guidelines on Quality Criteria for Diagnostic Radiographic Images in Paediatrics, EUR 16261, ISBN 92-827-7843-6, 1996.
7. European Guidelines on Quality Criteria for Computed Tomography, EUR 16262, ISBN 92-828-7478-8, 1999.
8. CT Safety and Efficacy, A Broad Perspective, www.msct.eu/CT_Quality_Criteria.htm.
9. Council Directive 97/43/Euratom, Official Journal L 180, pp. 22–27, 1997.
10. Guidelines on Education and Training in Radiation Protection for Medical Exposures, http://ec.europa.eu/energy/nuclear/radiation_protection/doc/publication/116.pdf, 2000.
11. Shiralkar et al., Br. Med. J., 327, 2003.

Installation Shielding in Cyclotron Facilities

Riccardo Calandrino

Abstract The topics of this lecture will include:

1. Elemental Activation.
   1.1 Nuclear reactions and activation.
   1.2 Activation of the beam line component.
   1.3 Air Activation Reactions.
2. Medical Cyclotrons.
   2.a Self shielded Systems.
      2.a.1 Ventilation and air changes.
      2.a.2 Safety.
   2.b Open Systems.
      2.b.1 Shielding Calulation guidelines.
      2.b.2 Ventilation and air changes: rules and references.
      2.b.3 Safety.
3. Cyclotron Shielding Decommissioning.

Keywords Cyclotron • Radioprotection • Activation • Shielding calculations • Decommissioning

1 Elemental Activation

1.1 Nuclear Reactions and Activation

Activation phenomena are generated by beam particles or by secondary particles interacting with beam line elements and the surroundings. The main parameter describing the probability of the interaction is the Cross Section Value, a dimensional

R. Calandrino (✉)
Medical Physics Department IRCCS San Raffaele, Milano, Italy
e-mail: calandrino.riccardo@hsr.it

Y. Lemoigne and A. Caner (eds.), *Radiation Protection in Medical Physics*, NATO Science for Peace and Security Series B: Physics and Biophysics 1, DOI 10.1007/978-94-007-0247-9_16,

parameter indicating a surface. Its unit is the barn – 10^{-24} cm^2; the mean nucleus radius is consequently 10^{-12} cm.

When a given material with N nuclei over cubic centimetres is bombarded with I particles over (s × cm^2), assuming that each nucleus has a mean cross section, in the direction of the beam equal to **σ**, we obtain the number of reactions/(s × cm^3), as follows:

$$IN\sigma$$

σ (E) represents the function indicating the probability for a given reaction. It is a function of the energy of the beam particle and of the mean section of the nuclei. Therefore we can have reactions with $\sigma = 0$ when the energy of the particle is below the threshold, as well as reactions with σ greater than the geometrical cross section of the target nucleus, as for example the cross section for thermal neutron capture of Cadmium (2,450 barns).

Considered Reactions:

- Proton beam induced reactions
- Secondary neutron induced reactions

1.1.1 Activation of Beam Line Components (Proton Beam Induced Reactions)

In this paragraph we shall deal with all the reactions induced by protons. The target for these reactions will be the following cyclotron components:

- Beam collimation systems
- Target (content)
- Target (container)

The alloys for various cyclotron components and the level of their induced activations are reported in Table 1.

It is evident that the highest activation values are found in the parts localized at the end of the accelerating path of the protons: The diaphragms of separation (window and vacuum foils) between the inner target volume and the vacuum volume of the cyclotrons, normally called the dees sector, where protons are continuously accelerated, will be the hottest parts.

Stripping forks (the holders of the stripping foils) and carousels could also present noticeable levels of activation. The level of activation of the foils at EOB could be of the order of 10^9 Bq, with a dose rate in the proximity of the part of up to 10–20 mSv/h.

The rule requiring the postponement of maintenance for at least 24 h from the last bombardment lowers the dose rate by a factor of 100, allowing the staff to perform the needed adjustments.

Targets are currently made of Niobium, which activates producing Mo93 and Mo91, two short lived elements.

Table 1 Alloys of cyclotron components and their induced activation

Part	Made of (IBA 18 MeV)	Made of (CTI 11 MeV)	Risk for activated particulate generation
Ion source anode	Cu, W	Not known	Very low
Ion source cathodes	Ta	Ta	Very low
Dees	Cu	Cu	Very low
Accelerating chamber (vacuum chamber)	Al, Mg, Si (Fe, Ni, Mn)	Al	Very low
Stripping forks and carousels	Al, Mg, Si (Fe, Ni, Mn)	Graphite (C)	Medium for IBA cyclotron, very low for CTI cyclotron
Stripping foils	C	C	Very low
Collimator	Aluminium	Aluminium	Very low
Window Foils	Havar, aluminium	Havar	Very high
Vacuum foils	Havar, aluminium, titanium	Aluminium	Very high
F-18 Targets	Silver, niobium, aluminium	Silver	Low
C-11 targets	Aluminium	Aluminium	Low

Table 2 Main reactions induced by neutrons flux

Reaction	T/2	Threshold (MeV)	Cross section
$O^{16}(n, p)N^{16}$	7.5 s	10	0.04 b
O^{16} (n, 2n)0^{15}	2 min	18	0.02 b
$N^{14}(n, p)C^{14}$	5,730 years	0, 5	0.10 b
N^{14} (n, p)C^{14}	5,730 years	Catt.term.	1.81 b
N^{14} (n, t)C^{12}	12 years	4, 3	0.02 b
N^{14} (n, 2n)N^{13}	10 min	11, 3	0.01 b
$Ar^{40}(n, p)Cl^{40}$	1,4 min	6, 9	0.01 b
Ar ^{40} (n, np)Cl^{39}	55 min	10, 2	0.001 b
Ar ^{40} (n, d)Cl^{39}	55 min	12, 4	0.001 b
Ar ^{40} (n, α)S^{37}	5.0 min	2, 6	0.001 b
Ar ^{40} (n, γ)Ar^{41}	1.83 h	Thermal	0.5

As a final judgement we consider the following list to represent the most likely isotope production by activated cyclotron parts:

$$\text{Tc}-97,\ \text{Co}-56,\ \text{Co}-57,\ \text{Co}-58,\ \text{V}-49,\ \text{Fe}-55$$

and with lower probability (or specific activity)

$$\text{Cd}-109,\ \text{Zn}-65 \text{ and } \text{Na}-22$$

1.1.2 Air Activation Reactions (Neutrons Induced)

The neutron flux induces activation in the elements of the air mixture. The main reactions are listed in the Table 2.

2 Medical Cyclotrons

2.1 *Self Shielded Systems*

Self shielded Cyclotrons are convenient in case of limits in the space for additional shieldings. For cyclotrons with beam energies between 15–20 MeV, 180–220 cm concrete is needed, while a self shielded cyclotron is directly provided of composite shield made of Pb+ borated poly with a total thickness of 120–140 cm.

Advantages of the self shielded cyclotron:

- Volume reduction for vault shieldings
- Decrease in the safety systems for vault control as consequence of the "low" internal dose rate
- No air activation induced problems

Disadvantages of the self shielded Cyclotron:

- High cost
- High accuracy requirements for floor level precision, and its stability over time

2.1.1 Ventilation and Air Changes

No problem concerning the outlet of air activated elements: air volume inside the shields is of the order of a few litres

In consideration of the risk of leakage of the target during the irradiation, it is recommended to consider the cyclotron vault, following UNI 10491(whose main definitions are reported in Table 3). as area C, and therefore the air changes and the depressure values as follows:

- Depressure 50–100 Pa
- Air volume changes: 5–10 vol/h (suggested >10)

2.1.2 Safety

The suggested safety systems are:

- Shieldings – two switches for the beam on allowance
- Door – two switches for the beam on allowance
- Vault-double level control
- Radiation dose rate Plastic Scintillator
- Air contamination monitoring by continuous sampling

Table 3 UNI 10491

Zone	Max dose (mSv/y)	Air changes/h	Gradient of depressure (Pa)
A	<6	2–5	10–20
B	<20	2–5	30–60
C	>20	5–10	50–100
D	>20	10–20	150–450

2.2 *Open Systems*

2.2.1 Shielding Calculation Guidelines

The standard procedure for the shielding calculations is that indicated by NCRP 51. For a p, n reaction on Al (target) the neutron fluence for uAmp is assumed, conservatively, to be ([1] appendix F Fig. F1) (Fig. 1):

$$\phi_0 I^{-1} = 10^6\, neutrons\, cm^{-2}\, \mathrm{s}^{-1} m^2 \mu A^{-1}$$

For example, for a mean current value of 80 μAmp, we obtain:

$$\varphi_0 = 8 \bullet 10^7 \text{ neutrons cm}^{-2}\text{s}^{-1}\text{m}^2$$

Assuming the conversion factor defined by the reference [1], we can finally construct a reference table for the barrier thicknesses at different distances from the source point and for different allowable weekly maximum dose rates (Table 4):
Where:

Rt is the dose rate at reference point without any barrier in cSv/week,
D1 is the maximum allowable dose rate in μSv/week,
Cr is the ratio between D1 and Rt,
Sc is the thickness of the barrier in cm of concrete.

2.2.2 Ventilation and Air Changes: Rules and References

The reference Guidelines are found in the publication UNI 10491.

The Cyclotron room for UNI 10491 is to be considered area type D; the ventilation requirements are:

- Depressure 150 Pa
- Number of air changes > 1–20 vol/h

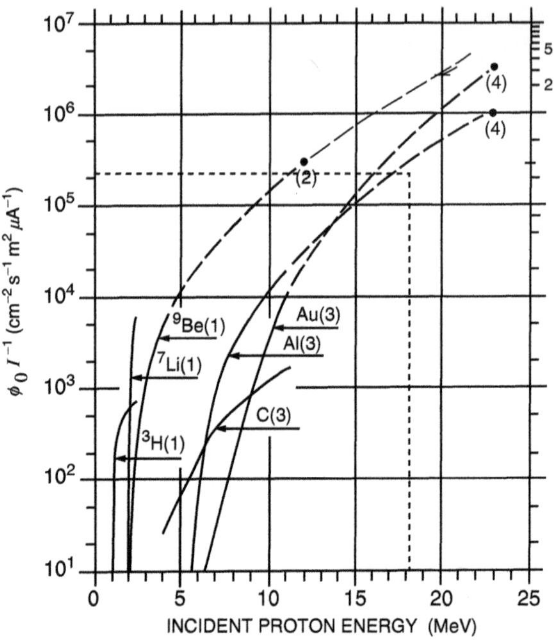

Fig. 1 Thick-target neutron fluence rates for (p, n) reactions ([1] Appendix F)

Table 4 Barrier thickness versus distance from the source and different dose rates

Distance (m)	Φ	Rt (cSv/week)	D1 (uSv/week)	Cr	Sc
1	8.00E + 07	4.03E + 05	10	2.48E–09	258
1	8.00E + 07	4.03E + 05	40	9.92E–09	240
1	8.00E + 07	4.03E + 05	400	9.92E–08	210
2	2.00E + 07	1.01E + 05	10	9.92E–09	240
2	2.00E + 07	1.01E + 05	40	3.97E–08	222
2	2.00E + 07	1.01E + 05	400	3.97E–07	192
3	8.89E + 06	4.48E + 04	10	2.23E–08	230
3	8.89E + 06	4.48E + 04	40	8.93E–08	211
3	8.89E + 06	4.48E + 04	400	8.93E–07	181

2.2.3 Safety

Door

Door opening not allowed when:

- Beam ON
- Room dose rate > 100 μSv/h
- The door must not be closed if:
- Round not completed
- More than 60 s has passed since last round button pushed.

Cyclotron is not allowed to Beam ON, when:

- No signal from air conditioning
- Rounds not completed
- Room door opened
- Emergency ON pushed

3 Cyclotron Shielding Decommissioning

The secondary neutron flux in the course of time generates activated elements in the components of the perimetral walls. The activation values are a function of the energy of the beam and of the workload of the facility. The levels of activation of the shields, the floor and perimetral walls of an 11 MeV recently decommissioned self shielded cyclotron were calculated using a Monte Carlo Code, and measured by gamma spectroscopy [2].

The analyzed reactions were: (n, γ), (n, p), (n, α), (n, d), (n, t) and (n, 2n), with a mean neutron energy of 2 MeV. The considered reactions yielding radioactive isotopes with half life greater than 27 days are listed in Table 5.

Specific activity simulated data averaged on 80 cm with a 10 cm step, from the surface of shields and floor are listed in Table 6. The values obtained from sample measurements for both laboratories were also reported for gamma emitting isotopes.

Table 5 Reactions yielding isotopes with half-life >27 days

Cyclotron component	Nuclide	Reaction	Activated elements	Half-life
Shields and floor				
	O-17	(n, α)–>	C-14	(5.730e3 a)
	Ca-40	(n, γ)–>	Ca-41	(103.0e3 a)
	Ca-40	(n, p)–>	K-40	(1.277e9 a)
	Ca-42	(n, α)–>	Ar-39	(269.0 a)
	Ca-44	(n, γ)–>	Ca-45	(162.7 d)
	Fe-54	(n, γ)–>	Fe-55	(2.700 a)
	Fe-58	(n, γ)–>	Fe-59	(44.64 d)
	Eu-151	(n, γ)–>	Eu-152	(13.60 a)
	Eu-153	(n, γ)–>	Eu-154	(8.800 a)
	Co-59	(n, γ)–>	Co-60	(5.271 a)
	S-33	(n, p)–>	P-33	(25.40 d)
	S-34	(n, γ)–>	S-35	(87.45 d)
	K-39	(n, γ)–>	K-40	(1.277e9 a)
	K-39	(n, p)–>	Ar-39(1	(269.0 a)
	K-39	(n, α)–>	Cl-36	(301.0e3 a)
	H-2	(n, γ)–>	H-3	(12.28 a)

Table 6 Calculated and measured induced activation on shields and floor

Isotope	Decay	Specific activity (Bq/kg) Simulated mean on 80 cm	Simulated @ 10 cm	Measured lab1 @ 10 cm	Measured lab2 @10 cm
Shields					
C-14	EC	1.63E + 00	1.11E + 00		
Ca-41	EC	2.34E + 00	1.58E + 01		
Ar-39	β–	2.80E + 00	2.17E + 01		
Ca-45	β–	8.05E + 02	5.45E + 03		
Fe-55	EC	2.45E + 03	1.66E + 04		
Fe-59	β–	7.62E + 00	5.18E + 01		
Eu-152	EC; β+ ; β–	1.32E + 02	8.95E + 02	3.6E + 01	3.0E + 01
Eu-154	EC; β–	7.56E + 00	5.21E + 01		
Co-60	β–	4.80E + 01	3.28E + 02	2.2E + 01	2.1E + 01
S-35	β–	4.56E + 00	3.08E + 01		
Mn-54				5.1E + 00	5.2E + 00
Others		1.67E – 01	8.28E – 01		
Floor					
Ar-39	β–	2.96E + 00	2.29E + 01		
Ca-45	β–	3.11E + 01	2.13E + 02		
Fe-55	EC	9.51E + 01	6.54E + 02		
Fe-59	β–	3.16E – 01	2.20E + 00		
Eu-152	EC; β+ ; β–	4.75E + 00	3.21E + 01	1.4E + 02	1.5E + 02
Eu-154	EC; β–	3.06E – 01	2.14E + 00	1.4E + 01	1.9E + 01
Co-60	β–	1.90E + 00	1.32E + 01	7.3E + 01	8.4E + 01
S-35	β–	1.66E – 01	1.13E + 00		
Sc-46				2.6E + 01	2.9E + 01
Mn-54				3.5E + 00	2.2E + 00
Others		1.18E – 01	8.83E – 01		

The mismatching between calculated and measured data is due to the difficulty of providing samples in the exact position considered for simulation. In addition, gamma-ray spectrometry has identified the presence of Mn-54 and Sc-46 both in shields and floor. Mn-54 is produced by (n, p) reaction on Fe-54, impurity of concrete (8.7×10^{20} atoms cm^{-3}) [3] [6]. The isotope was not considered in the simulation due to the very low cross section of the reaction (0.386 barn), but the abundance of Fe-54 in the concrete justifies the experimental data. The presence of Sc isotope is evidently due to neutron capture on trace amounts of Sc-45 (3.9×10^{17} atoms cm^{-3}) that was not included in the elemental contents of concrete.

The simulated specific activity (stratified on 10 cm thick slabs) was utilized to evaluate the radiation rates in the proximity of the different components expected to be activated, giving a preliminary assessment of the risk for the staff involved in the decommissioning. The results are compared with the measured values in Table 7.

Target activation has not been considered in the simulation because of the limited volume of this component. In any case, these are the hottest components and their rates are readily measurable.

Dose rates originating from various targets are summarized in Table 8. Obviously the dose rate is a function of both beam time and isotope production.

The total activity for each component has been evaluated as reported in Table 9; these values have been considered for the legal allowances in our country.

Table 7 Simulated and measured dose rate

Measure distance	Component	Dose rate calculated (mSv/h)	Dose rate measured (mSv/h)
Contact	Shields	2.41E−07	0.0

Table 8 Dose rates originated by various targets

Target	Dose rate at 5 cm (μSv/h)	Dose rate at 30 cm (μSv/h)	Working time
O15	0.20	0.20	Low
N13	0.20	0.20	Low
C11	0.90	0.35	Low
F18	16.0	3.0	High

Table 9 Total activity per component

Part	Total activity (Bq)
Shields	1.71E + 07
Magnet	2.92E + 08
Target 4	3.38E + 05

As a general conclusion, the decommissioning of a self-shielded cyclotron of 11 MeV after a 15 year working life does not represent a risk for staff involved in its dismantling when continuously supervised by trained professionals.

The survey and monitoring of the environment and of the workers during the dismantling phase by the Health Physics Staff and by qualified Spectroscopy Labs represent a requisite condition for the safe outcome of the procedure.

The presence of long-lived contaminants in different parts of the accelerator and of the shieldings must be taken into consideration for the classification of the waste disposal in accordance with the relevant regulations and national laws.

In Italy the limits and legal prescriptions oblige the customer to store activated parts indefinitely in authorized areas for radioactive waste disposal.

This aspect should be considered in any facility prior to a decision regarding cyclotron installation.

References

1. NCRP 51 Radiation Protection design guidelines for 0,1 – 100 MeV particle accelerator Facilities; 1977.
2. R. Calandrino, A. del Vecchio, A. Savi, S. Todde, V. Griffoni, S. Brambilla, R. Parisi, G. Simone, F. Fazio - Decommissioning procedures for an 11 MeV Self Shielded Medical Cyclotron after 16 years working time. Health Physics, 588–596, 2006.
3. UNI 10491 Criteri per la costruzione di installazioni adibite alla manipolazione di sorgenti radioattive non sigillate; nov 1995.

Shielding and Facility Design in Nuclear Medicine

Renato Padovani

1 Introduction

The design and operation of a nuclear medicine department should combine the general principles of protection of the worker, the patient and the general public maintaining the characteristics of an efficient medical department.

This lecture discusses requirements for the design and organization of a nuclear medicine facility where radionuclides are used. The presentation deals solely with requirements to ensure radiological safety. Other occupational health and safety considerations, such as control of bio-hazardous material or reduction of fire hazards, may require adherence to additional guidelines.

2 Categorisation of Hazard

ICRP Report 57 suggests a method to determine the broad requirements for planning any particular clinical nuclear medicine facility. The first step is to calculate the weighted activity. To obtain the weighted activity, the largest activity likely to be encountered at any time in the area to be planned must be determined. This figure is multiplied by a modifying factor according to the radionuclide being used (Table 1). The weighted activity-modifying factor takes into account the radionuclide's radiotoxicity (emission type and energy, half-life, bio-distribution, etc.). Thus ^{3}H would have a lower weighted activity, while radionuclides such as ^{125}I or high-energy beta emitters such as ^{89}Sr and ^{32}P would be accorded a higher weighted activity.

The weighted activity is then multiplied by a second modifying factor (Table 2) determined by the nature of the operation. This takes account of the greater hazards of complex radiopharmaceutical preparation and of the lower hazards associated with storage and activity administered to patients.

R. Padovani (✉)
Medical Physics Department, University Hospital S. Maria della Misericordia, Udine, Italy
e-mail: padovani.renato@aoud.sanita.fvg.it

Y. Lemoigne and A. Caner (eds.), *Radiation Protection in Medical Physics*,
NATO Science for Peace and Security Series B: Physics and Biophysics 1,
DOI 10.1007/978-94-007-0247-9_17, © Springer Science+Business Media B.V. 2011

Table 1 Weighting factors according to radionuclide

Class	Radionuclide	Radiotoxicity weighting factor
A	^{75}Se, ^{89}Sr, ^{125}I, ^{131}I, ^{32}P, ^{90}Y, ^{99}Mo, ^{153}Sm	100
B	^{11}C, ^{13}N, ^{15}O, ^{18}F, ^{51}Cr, ^{67}Ga, ^{99m}Tc, ^{111}In, ^{113m}In, ^{123}I, ^{201}TI	1.0
C	^{3}H, ^{14}C, ^{81m}Kr, ^{127}Xe, ^{133}Xe	0.01

Table 2 Weighting factors according to type of operation

Type of operation or area	Operational weighting factor
Storage	0.01
Waste handling	
Scintigraphic counting/imaging when administration is made elsewhere	0.1
Patient waiting area	
Patient bed area (diagnostic)	
Local dispensing	
Radionuclide administration	1.0
Scintigraphic counting/imaging when administration is made in same room	
Radiopharmaceutical preparation, simple	
Patient bed area (therapeutic)	
Radiopharmaceutical preparation, complex	10.0

Table 3 Categorisation of hazard

Weighted activity	Category
Less than 50 MBq	Low hazard
50–50,000 MBq	Medium hazard
Greater than 50,000 MBq	High hazard

The final figure obtained after the two steps above gives the weighted activity. The category of hazard to workers can then be determined from Table 3.

Once the category of hazard has been determined, the main requirements of the facility can then be determined from Table 4.

3 Planning and Localisation

When the nuclear medicine is part of a hospital, the location should be such that there is no interference from other radiation emitting equipment, such as X-ray units, teletherapy units (Co-60, linear accelerator) etc. Normally, it is preferred that the nuclear medicine department is located at the end of a hospital block where movement of public is restricted. At the same time it should be well connected with other departments.

The design of the facility should take into consideration the type of work and the radionuclides and their activities intended to be used. The concept of 'categorization

Table 4 Facilities required for radiation protection in relation to category of hazard

Category of hazard	Structural Shielding	Floor	Surfaces	Fume hood	Ventilation	Plumbing	First aid
Low	Nil	Cleanable	Cleanable	No	Normal	Standard	Washing facilities
Medium	Nil	Continuous	Cleanable	Yes	Good	Standard	Washing and decontamination
High	Possibly	Continuous	Cleanable	Yes	May need special forced ventilation	May need special plumbing facilities	Washing and decontamination facilities

of hazard' should be used in order to determine the special needs concerning ventilation, shielding, materials used in walls, floors and work benches.

The medical physicist with his/her role as radiation protection adviser (RPA) should be consulted as soon as the planning process commences for construction or renovation of a nuclear medicine facility or other hospital radioisotope laboratory. Close collaboration with senior personnel who will be operating the laboratory and the architect or project manager is essential. The laboratory operator, RPA, planners and senior management should also be aware that with the rapid evolution of clinical and research work involving radioisotopes, it may be prudent to build more radiological protection features into a laboratory so that it would be suitable for almost any type of radioisotope work in the long term (20–30 years). Incorporating desirable features is usually far simpler and less expensive than subsequent refitting.

Care should be taken to ensure that the safety requirements necessary for radioisotope use do not compromise the safety requirements for the use of other hazardous agents (infectious, anti-neoplastic, chemical, etc).

4 General Layout Requirements

The general layout of a department for the use of unsealed sources can be described under three separate headings:

1. Laboratories and premises not frequented by patients, including rooms for the storage, preparation, and dispensing of radiopharmaceuticals including radioactive waste
2. Areas occupied by patients, including rooms for the administration of radioactive materials and for carrying out measurements on patients
3. Offices

The basic principle in the general layout should be to separate high activity areas from low activity areas and separate working areas from areas occupied by the patients. Offices and reception should be located in the low activity area.

The Licensee should consider access control when determining source storage areas and rooms for hospitalized patients undergoing radionuclide therapy.

Regular contamination monitoring should be made with a minimum frequency according to the requirements of the Regulatory Authority.

The different rooms of the facility should be used only for the intended work.

4.1 Laboratory and Premises Not Frequented by Patients

The importance and the number of these rooms vary widely in different hospitals. According to the workload, the number of radionuclides and activities used, and the type of work performed, it is desirable to have separate rooms for the main types of work and radionuclides.

4.1.1 Rooms for Preparation and Dispensing of Radiopharmaceuticals (Hot Lab)

Most handling of radiopharmaceuticals requires the use of aseptic techniques. Special requirements for radiation protection and for hygiene may be required.

Such considerations require installation of special ventilation systems, including fume hoods and laminar-air-flow cabinets, and these must be regularly controlled with regard to airflow velocity and the effectiveness of the filter.

All premises should be regularly monitored for contamination. Equipment for continuous monitoring of external exposure should be considered in rooms aimed for preparation of radiopharmaceuticals. Monitors and their warning devices should always be calibrated and operability should be checked prior to each day of use.

A removable coating on the floor will reduce the chances of radioactive material being ground into the flooring and simplify any decontamination procedure. The floor covering must be smoothly coved (curved) to the walls and benches, and should be glued to the floor to prevent liquids spreading underneath in the event the covering is punctured. Care should be taken to seal all joints to minimize the trapping of radioactive material.

Walls should be painted with washable, non-porous paint.

It is unlikely that shielding in the walls will be necessary in low-level laboratories, since localized shielding is usually adequate; however, the need for wall shielding should be assessed in medium- and high-level laboratories.

Worktop surfaces must be finished in a smooth, washable and chemical-resistant surface with all joints sealed. It should be borne in mind that structural reinforcement may be necessary, since a considerable weight of lead shielding may be placed on counter tops.

Laboratories in which radioactive aerosols or gases may be produced or handled should have an appropriate ventilation system that includes a fume hood or glove box. The ventilation system should be designed such that the laboratory is at negative pressure relative to surrounding areas. The airflow should be from areas of minimal likelihood of airborne contamination to areas where such contamination is likely. All air from the laboratory should be vented through a fume hood and must not be re-circulated either directly, in combination with incoming fresh air in a mixing system, or indirectly, as a result of proximity of the exhaust to a fresh air intake.

Washing facilities including a shower should be available as well as a shower for cleaning the eyes in case of contamination.

Work benches should be covered with absorbent paper with a layer of plastic on one side, in order to absorb and contain spilled liquids at once.

A bench top shield with a lead glass window should always be used. Vial shields and syringe shields for the different radionuclides and different sizes of vials and syringes should be available.

A shielded container for used syringes, needles and vials as well as a pedal-operated bucket for other types of radioactive waste such as used gloves and contaminated paper should be used.

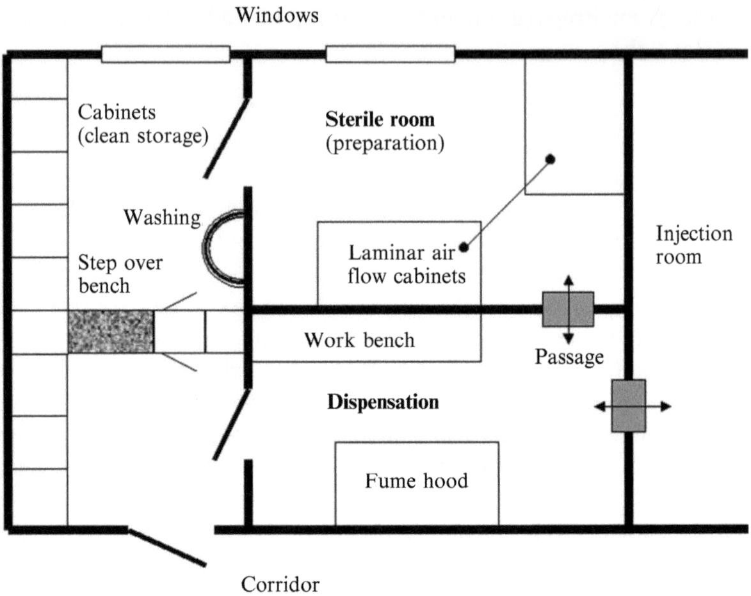

Fig. 1 Example of rooms for preparation and dispensation of radiopharmaceuticals

Instruments for remote handling of the radionuclides should be available and must always be used when transferring an unprotected vial or a syringe, for instance, to the activity meter.

Protective clothing should always be used when working with radioactive material. Instruments, equipment and material to check for contamination and for decontamination procedures should be available.

A sign should be posted on the door warning of the radiation hazard (Fig. 1).

4.1.2 Facility for Storage of Radionuclides

All radioactive materials must be stored in a secure location to prevent unauthorized access to the material.

The majority of radionuclides used in nuclear medicine procedures have relatively short half-lives, usually between 6 h and a month. Longer-lived radionuclides are more commonly used in in-vitro procedures and in biomedical research laboratories.

Storage, initial opening of vials and dispensing of radioisotopes (as received from the supplier) must be carried out in a designated radiation work area equipped with absorbent bench covering material.

Locks must restrict access to areas where radioactive materials are used and stored and only authorized personnel should have access. Medium- and high-level radioisotope laboratories and waste storage repositories require greater security and are usually required to have a good lock on the door(s) and may also be provided with lockable storage areas within the facility.

4.1.3 Liquid and Solid Waste

The Regulatory Authority should specify whether or not it is acceptable to dispose of aqueous radioactive waste directly into the sewer. If it is permissible, drain pipes from the radioisotope laboratory sink should go as directly as possible to the main building sewer, and should not connect with other drains within the building, unless those other drains also carry radioactive material. This is to minimize the possibility of a "back up" contaminating other, non-controlled areas. The final plans of the drainage system which are supplied to maintenance personnel must show which drains are from radioisotope laboratories.

A conventional stainless steel sink is more commonly used for washing. The sink drain trap should be accessible for monitoring for contamination build up.

As for solid wastes, it is important to keep the short-half-life material separated from long-lived radionuclide waste. Short half-life radionuclide waste from ^{99m}Tc ($T_{1/2}$ = 6 h) will decay to background levels within 3 days, and is usually stored in-house and then disposed of as non-radioactive waste. The appropriate Regulatory Authority will usually indicate the special requirements for the collection and disposal of longer-lived radionuclide waste.

Medium- and high-level radioisotope waste storage repositories require for security to have a good lock on the door(s) and may also be provided with lockable storage areas within the facility.

4.2 Areas Occupied By Patients

Any room where radiopharmaceuticals are administered to patients should be designed as a medium or high level laboratory because of possible contamination.

The injection room should be close to the radiopharmacy (hot laboratory) to minimize the movement of radiopharmaceuticals.

Segregation of injected patients in a separate waiting room is recommended. While injected patients are a source of radiation, the dose rate in their immediate environment is not usually high enough to give a significant dose to other patients or medical staff sharing the waiting room on an infrequent or casual basis. However, the patient waiting area should not be sited immediately adjacent to desk areas for receptionists or other workers as the accumulated dose in close proximity over a year can easily exceed 1 mSv. Consideration should therefore be given to locating this waiting room in a low occupancy area, or incorporating some shielding into walls or partitions to minimize accumulated doses to workers at nearby desk stations.

A separate toilet room for the exclusive use of injected patients is also recommended. A sign requesting patients to flush the toilet well and wash their hands should be displayed to ensure adequate dilution of excreted radioactive materials and minimise contamination. Washrooms designated for use by nuclear medicine patients should be finished in materials that are easily decontaminated.

There should be ample space not only for the nuclear medicine staff and patients, but also for the essential equipment, accessories and supplies used in the procedures. Gamma cameras should be separated as much as possible, either by distance or shielding partitions. This not only reduces doses to technologists working at the camera, but also reduces the possibility of extraneous radiation being received by the camera as a result of activity in nearby patients. Space should be such that technologists can retreat at least 1 m, preferably 2 m, from injected patients not requiring close supervision. Alternatively, mobile barriers might be used to shield the technologist.

5 Final Plan and Classification of Areas

A plan of a small nuclear medicine facility is shown in Fig. 2. This is intended only to illustrate some of the principles discussed above.

As mentioned above, the National and International Regulations require that the different areas in a nuclear medicine department should be classified as controlled, supervised and non-controlled. The radiation protection committee and the radiation protection officer should do the classification. The following rooms are suggested to be classified as controlled areas:

- Rooms for preparation, dispensation and storage of radiopharmaceuticals
- Rooms for administration of radiopharmaceuticals
- Room for temporary storage of radioactive waste
- Examination rooms
- Other laboratories for preparation and measurements of radioactive samples

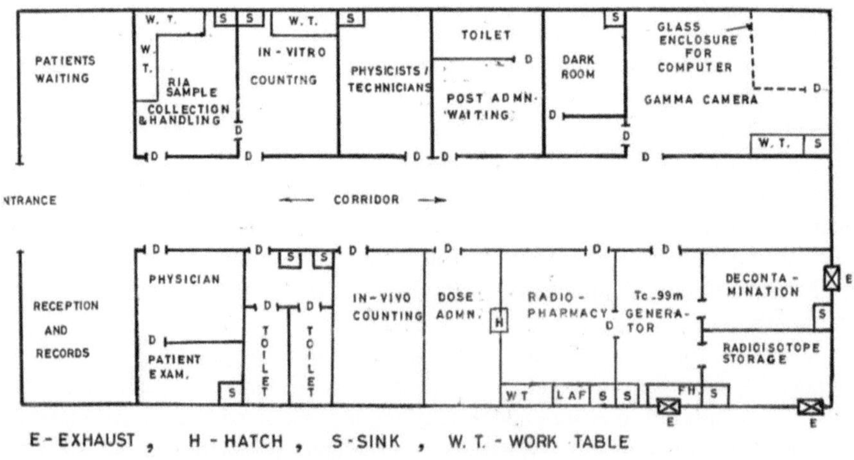

Fig. 2 Example of a plan of a small nuclear medicine department

The following rooms should be classified as supervised areas:

- Waiting room and toilet for radioactive patients
- Laboratories for Radioimmunoassay analysis (RIA)
- Other areas such as the reception, offices etc. do not need any classification

References

1. Radiological protection of the worker in medicine and dentistry, ICRP Publication No. 57. Oxford, Pergamon Press (1989) (Annals of the ICRP 20, 3).
2. WHO, Manual on radiation protection in hospitals and general practice, Volume 1: basic Requirements and Volume 4: Nuclear Medicine (Manuscript 2000).
3. IAEA, Regulations for the safe transport of radioactive material Safety Standards Series No. ST-1/ Requirements (1996).
4. IAEA, A Safety Guide on Predisposal management of radioactive waste from medicine, industry and research, Safety Standard Series No 111-G-2.2 (1997).

Installation Shielding in Radiotherapy

Riccardo Calandrino

Abstract The topics of this lecture will include:

General
  ICRP parameters W, U, T
Part I: Linac installation
  Standard layout for the radiotherapy room
  Primary and secondary radiation
  Ever green NCRP51
  Area limits
  NCRP151 approach to calculations
  Neutron contamination of high energy units
  Interactions of neutrons with matter
  Door calculation
Part II: Brachytherapy Rooms
  Sources and their physical characteristics
  Two examples: HDR and LDR shielding calculations

Keywords Radiation protection • Linac shielding • Brachytherapy shielding • TVL

1 Introduction: Design of the Radiotherapy Room

The purpose of radiation protection is well defined by the ALARA principle. It indicates that a radiotherapy room with barriers of a mean thickness of 4 m or more, while surely safe, would also represent a waste of money and space without any reasonable improvement relative to radioprotection requirements.

R. Calandrino (✉)
Medical Physics Department IRCCS San Raffaele Milano, Italy
e-mail: calandrino.riccardo@hsr.it

Y. Lemoigne and A. Caner (eds.), *Radiation Protection in Medical Physics*, NATO Science for Peace and Security Series B: Physics and Biophysics 1,
DOI 10.1007/978-94-007-0247-9_18,

Nevertheless, an underestimation of the "safety" thickness of the barriers could lead to dangerous exposure of the staff and population. Therefore it is highly recommended to acquire the most advanced information concerning the calculation methods and the shielding efficiency of the different materials before initiating the design of any radiotherapy room.

2 Linac Installation

2.1 Primary Radiation Shielding

The primary radiation output of a Linac is calculated on the basis of the Work Load Factor: W.

The Use Factor of the wall U, multiplied by the work load W, and divided by the square of the distance from isocenter + 1 (distance from the source), will give the radiation dose rate at the reference point.

The Occupancy factor of the area (T) related to the various area limits, defined by ICRP60 [1] and\or ICRP 103 [2], given in Table 1, will define the value for the maximum allowable rate at the reference point outside the barriers.

For example: for a corridor adjacent to the lateral wall of a treatment room, we shall accept, as the maximum allowable rate:

$$5\,\mu\text{Sv/week} * 1/10 = 50\,\mu\text{Sv/week} \tag{1}$$

where the first factor comes from the data reported in Table 1 for free areas, and 1/10 is the use factor commonly used for corridors, waiting rooms and bathrooms.

The ratio between the maximum allowable rate and the rate, unshielded, at the reference point, will give the required attenuation coefficient.

For example: calculating the required barrier for a lateral wall where the reference point is 6 m from the isocenter, the formula will be:

$$1{,}000 * 0.25 * (1/7)^2 = 5.10\ \text{Gy/week} \tag{2}$$

where 1,000 Gy/week is the assumed workload of the facility, 0.25 is the use factor for the lateral wall and the third term is the reduction due to the inverse square law.

Table 1 Occupancy factor of area versus area limits

Classified area	Max rate (μSv/week)	Project rate (μSv/week)
Controlled	400	100
Surveilled	100	30
Free	20	5

The ratio between the data calculated at 1 and 2 is:

$$50/\left(5.10*10^{6}\right)=9.8\times10^{-6} =: \mathrm{Cr}\,(\text{attenuation coefficient})$$

The thickness of the barrier will be defined by the formula:

$$\mathrm{TVL}*(-\log \mathrm{Cr})$$

where TVL is the assumed value of the Tenth Value Layer for the radiation considered. (45.5 cm concrete in case of 18 MeV X rays).

The final result in this case will be:

$$45.5*5.01=228 \text{ cm concrete}$$

2.2 *Secondary Radiation Shielding*

Outside the areas exposed to the primary beam, the secondary radiation becomes the object of our calculations. Obviously the secondary radiation component exists also in the primary region, but the additional contribution, even if considered in the barrier thickness evaluation, has a negligible impact on the final value of the barrier thickness. As an example let us consider a standard calculation for a primary beam of an 18 Mev X as follows (Table 2).

In this case the final result for the required thickness is 239 cm of concrete.

If we cancel the secondary contribution as shown in Table 3, we obtain final difference of 1 cm between the two.

The rate of the secondary contribution derives from two basic contributions:

1. Accelerator scattered and leakage radiation (estimated between 0.1% and 0.5 %)
2. Patient outside scattered radiation (0.1 %)

Normally these are defined as scattered component (derived from patients and Linac scattered radiation), and leakage component.

Table 2 Shielding calculation for a primary beam

Variable	Quantity	Units	Definition
W	100,000.00	cGy/week	Weekly work load (1,000 Gy/week) primary
w′	600.00	cGy/week	Weekly work load (0,6% W) secondary
d	6.00		Distance between ref point and isocenter
U	0.25	Use factor	Use factor
R1	510.20	cGy/week	Primary rate at reference (cGy/Week)
R2	16.67	cGy/week	Secondary rate at reference (cGy/week)
Rt	526.87	cGy/week	R1 + R2
D1	30	uGy/week	Max allowable rate outside shielding
Cr	5.69E–06		Absorption
Sc	**239**	**cm concrete**	Barrier thickness

Table 3 Shielding calculation for a primary beam, excluding secondary contribution

Variable	Quantity	Units
W	100,000.00	cGy/week
w′	0.00	cGy/week
d	6.00	
U	0.25	
R1	510.20	cGy/week
R2	0.00	cGy/week
Rt	510.20	cGy/week
D1	30	uGy/week
Cr	5.88E–06	
Sc	**238**	**cm Concrete**

Table 4 Calculation of the addition of an additional HVL

Variable	Qauntity	Units	Definition
W	100,000.00	cGy/week	
wl	300.00	cGy/week	
ws	500.00	cGy/week	
d	9.00		
U	0.00		
R1	0.00		Primary rate
R2l	3.70	cGy/week	Leakage rate
R2s	6.17		Scatter rate
D1	40	uGy/week	
Crl	1.08E–03		Attenuation factor for the leakage component
Crs	6.48E–04		Attenuation factor for the scatter component
Scl	104	cm concrete	Thickness for leakage
Scs	92	cm concrete	Thickness for scatter
Sce	**115**	cm concrete	Final thickness for total seconds radiation

The scattered component has a lower energy relative to the primary beam; in fact the Compton scattered photon has lower energy than the generating primary photon, whereas the leakage component has the same primary energy.

To evaluate the correct thickness for the secondary barrier the two contributions must be consider separately, as suggested by NCRP 151, ultimately obtaining two different thicknesses.

In case the difference between the two values is less than 1 TVL of the leakage component, an additional HVL shall be added to the largest as follows (Table 4).

2.3 Neutron Transmission

For energies above 10 MeV a significant neutron production must be considered. This component does not represent a problem for the barrier so far calculated for the primary and secondary X rays, but it has to be considered when dealing with door and duct definition.

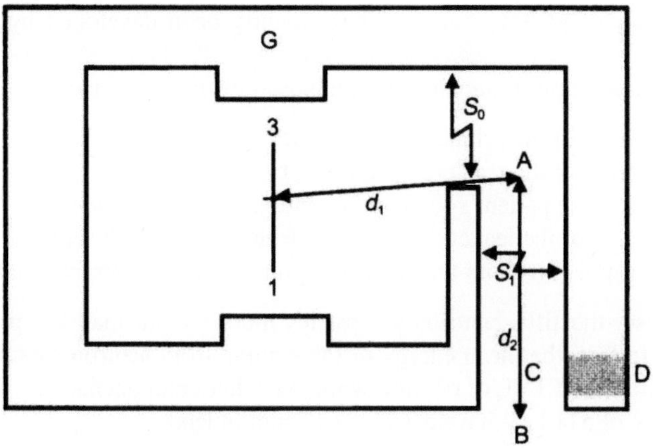

Fig. 1 Room layout for calculating neutron capture gamma-ray and neutron dose equivalents at the maze door

Table 5 Kersey method: typical layout installation data

Variable	Quantity	Units
S0	6.8	m^2
S1	8.2	m^2
d0	1.41	m
d1	5.5	m
d2	8.25	m
Hn, d	**1.94**	mSv/week
Thickness	158.98	mm polyethylene

The method to calculate transmission of neutrons along ducts has been definitively resolved by the NCRP publication 51. Various methods have been proposed in literature to calculate the neutron transmission along the entrance labyrinth of the treatment room. The Kersey method will be reported here.

We'll indicate the area of the inner part of the labyrinth aperture S_0 and the entrance corridor area S_1 (see Fig. 1).

Assuming a neutron contamination of the primary beam of 1.6 mSv/Gy at a distance of 1.4 m from the target (d0), with d1 and d2 defined as reported in the figure, the following data are obtained for a typical layout installation (Table 5).

2.4 Gamma Transmission

The transmission of gamma and\or X rays along the entrance maze is a matter of concern for the implications for thc correct assessment of the door and entrance corridor side wall thicknesses.

The theory of gamma transport has recently been developed by NCRP 151 ([4], 34–41).

The gamma flux is assumed to be composed of five components:

1. The scattering of the primary beam by the opposite wall
2. The scattering of the leakage component by the same wall
3. Scattering of the patient's scattered radiation by the same wall
4. Transmission of the leakage component from the labyrinth septum
5. Gamma originating from the neutron capture along the entrance maze

Among these, the fifth contribution carries more weight than the previous four combined. In fact, the mean energy of the gamma from neutron capture is around 3.6 Mev with a lead TVL of 61 mm, while the other components with an estimated mean energy of 511 keV have a TVL of 17 mm in lead.

Obviously this component can be ignored for energies ≤10 MeV of the X ray. In this case only components 1–4 of the above mentioned are to be considered.

This difference changes the door thickness for gamma absorption by at least a factor of 2. In fact, while 20 mm lead is the common solution for a 6 MeV unit, thicknesses up to 4–5 cm lead are needed for 15–18 Mev Units.

3 Brachytherapy Installations

The method to calculate the barriers for a Brachytherapy installation is very simple and will follow the same rules thus far presented.

The starting figure is the source strength (S_s) expressed in Bq. With the factor Γ (uGy * h^{-1}*MBq^{-1}*m^2) obtained from validated references, the rate at the reference point (at distance d) is calculated as follows:

$$Ss * \Gamma / d^2 =: \text{Rate at reference in } \mu\text{Gy h}^{-1}$$

Considering a mean time of 20 h/week, corresponding to the time when the source is outside the self-shielding unit, that is the Treatment ON time, (reasonable for an LDR, whereas 5 h/week is more appropriate for an HDR), the rate on a weekly basis can easily be calculated.

Once again the ratio between the maximum allowable rate and the above mentioned unshielded source rate yields the attenuation factor (Cr).

The formula:

$$\text{TVL} * (-\log \text{Cr})$$

will define the correct barrier thickness.

TVLs and Γs for the different isotopes used are shown in the following table [6] (Table 6).

Table 6 TVLs and Γs for the different used isotopes

Isotope	Mean energy (keV)	Half life	Gamma (μGy m2)/(h MBq)	TVL lead (mm)	TVL concrete (cm)
Co60	1250	5.3 years	0.306	42	28
Cs137	662	30.2 years	0.072	22	17.5
Ir192	380	74 days	0.116	16	14.7
I125	28	59.4 days	0.0337		
Pd103	21	17 days			

References

1. ICRP Publication 60; Ann ICRP 20(1) (1994) International commission on radiological protection. Recommendation ICRP. Oxford: Pergamon Press.
2. ICRP Publication 103 (2008) International commission on radiological protection. Recommendation ICRP. Oxford: Pergamon Press.
3. NCRP 51 (1977) Radiation protection design guidelines for 0, 1–100 MeV particle accelerator facilities.
4. NCRP 151 (2005) Structural shielding design and evaluation for megavoltage X and Gamma Ray radiotherapy facilities.
5. NCRP 79 (1984) Neutron contamination from medical electron accelerator.
6. Cristina Marchetti (2008) Organizzazione, progettazione e decommissioning di un impianto di Brachiterapia. AIFM Course: Bari Italy, Oct 2008.

Air Contamination Control in Radiochemistry Labs

Riccardo Calandrino

Abstract The topics of this lecture will include:

System to monitor air contamination in Radiochemistry Labs
Environmental safety: modelling of dispersion in the atmosphere of contaminated gasses and population dose estimation
Safety and shielding of Hot Cells
Internal contamination risk to staff during synthesis
Modelling for dose determination from urine sample measurements

Keywords Radiochemistry • Hot cells • Air contamination • Intake doses

1 Introduction

The enormous increase in the use of PET for diagnosis and staging in oncology has determined, over the past 5–10 years, a parallel demand for positron emitting radioisotopes and radiopharmaceuticals.

The most popular radionuclides are undoubtedly C-11 ($T_{1/2}$ = 20.3 min), which is usually obtained "in-target" as gaseous $[^{11}C]CO_2$, and F-18 ($T_{1/2}$ = 109.6 min), which may be generated both in gaseous ($[^{18}F]F_2$) and liquid ($[^{18}F]HF_{acq}$) forms. Both are routinely used in potentially complex radiosynthetic procedures in which one or more synthetic steps may lead to the formation of volatile radioactive by-products. The above processes are often performed by means of commercially available, automated synthesis modules. Typically, although not always, they include more or less effective trapping features, such as chemical absorbers or liquid nitrogen cooled glassware or coil traps. Furthermore, commercially available sealed

R. Calandrino (✉)
Medical Physics Department IRCCS San Raffaele, Milano, Italy
e-mail: calandrino.riccardo@hsr.it

Y. Lemoigne and A. Caner (eds.), *Radiation Protection in Medical Physics*,
NATO Science for Peace and Security Series B: Physics and Biophysics 1,
DOI 10.1007/978-94-007-0247-9_19,

hot cells are usually designed with outlet activated carbon filters which are generally incapable of effectively trapping the radioactive exhaust.

Unfortunately, the characterization and estimation of the radiation hazard generated by large scale production of PET radiopharmaceuticals is poor [1, 2]. The purpose of the first part of this lecture is to present the results of the outlet air contamination measurements obtained before and after the installation of an automated storage system, together with the evaluation of the doses to the surrounding population groups. In fact, the level of contamination of the samples, collected by the analyzer, can be as high as 1E05 Bq l^{-1} both during the C-11 labelled radiotracer preparations and the F-18 radiosynthesis in case of incomplete trapping.

The doses to the surrounding population, in terms of intake and external irradiation generated from the contaminated exhausted air, have been calculated for different production rates using a very simple model of conical dispersion, with the origin located at the top of the Hospital chimney. Calculations have been performed using the actual data obtained before and after the installation of the radioactive gas storage system. The results clearly demonstrate the need for such a system for facilities where the production of F-18 and/or C-11 above 5 Ci day^{-1} is routinely performed, in order to keep the dose to the surrounding groups of population below the limit of 10 μSv $year^{-1}$ [1].

The second part of the lecture concerns the risks and the doses for the internal contamination for the radiochemistry staff in a high workload medical cyclotron facility.

The doses from internal contamination derive from the inhalation of radioactive gas leaking out of the cells by the personnel involved in the synthesis processes. These are calculated from urine sample measurements. Different models will be presented to calculate the effective dose from the measurement of these urine samples and the results compared with data obtained from local environmental measurements of the released radioactivity inside the lab [2].

2 System to Monitor Air Contamination in Radiochemistry Labs

The air contamination monitoring system consists of a vacuum pump connected to a Marinelli glass beaker which is, in turn, connected to the source points through a series of flexible 30 mm diameter tubes that allow the transfer of air samples to the counting volume. The system is driven by software that selects the source point to be monitored by opening/closing the proper two-way valve following an operator programmed sequence of actions.

The Air Compressing Station consists basically of a compressor pump capable of concentrating the contaminated gas to a pressure of 200 bars into a series of eight cylinders located in a suitably shielded, dedicated room.

In our Hospital, three monitoring systems are in use to guarantee the surveillance of the radioactive dispersions in our facility where two cyclotrons and five Radiochemistry Labs are used for daily radioisotope production.

The radiochemists have been trained in the principles of the functioning of these systems and have the mandatory instruction to select the points to be sampled before initiating any radioisotope or radiopharmaceutical production.

As a general example, consider the following scenario. Two syntheses are simultaneously in progress in two different laboratories: C-11 in RC1 Lab of the Cyclotron 1 area, and [^{18}F]FDG in the RC3 lab, located in the cyclotron 2 area; the operators will then select the source points to be monitored as follows (highlighted).

System 1		System 2		System 3	
Measurement point	Status	Measurement point	Status	Measurement point	Status
Chimney	ON			Radioactive gas storage system	OFF
CTI cyclotron room	OFF	IBA cyclotron room	OFF		
Radiochemistry 1	ON	Radiochemistry 3	ON		
Radiochemistry 2	OFF	Radiochemistry 4	OFF		
Hot cells 1, 2	OFF			Hot cells 7, 8	ON
Hot cells 3, 4	ON			Hot cells 9, 10	OFF
Hot cells 5, 6	OFF				

The system is set to monitor every source point for 12 s, followed by a 30 second step needed to purge the Marinelli glass container, thus preparing it for the subsequent air sampling step. Summarizing, the cycle time for every measurement step is 0.7 min, while the overall time, supposing that all measurement points are set to the ON status, is 4.9 min for System 1 (which controls seven source points). This could potentially be in contrast with the need to carefully monitor processes whose time order is in the range of 5–10 min, as it is for the [^{11}C]CH_3I synthesis. This implies that a selection of the room/hot cells to be monitored, based on the daily working schedule, must be made.

For instance, in the above example, every point will be sampled every 1.4 min by monitoring system 1 and every 30 s by monitoring systems 2 or 3.

The ACS software continuously receives signals from both the air contamination monitoring systems and the Geiger-Mueller probes installed inside the hot cells. The program has been designed to activate the compressor every time two signals, sent by one of the G.M. hot cells and by the chimney monitor or internal Lab monitor, are simultaneously detected.

Monitoring system 1 is crucial, as it directly controls the quality of the air released into the atmosphere.

The different alarm thresholds are defined in order to keep the maximum release at a concentration value below 1 Bq/L following the rule, established by law, that the maximum allowable concentration in radioactive waste atmospheric disposal is, for elements with a $T_{1/2} < 75$ days , ≤1 Bq/gr and, considering the mean air density of 1.293 g/dm^3, it follows that the limit per litre of air is 1.293 Bq l^{-1}, rounded to 1.00 Bq l^{-1} [3].

In particular, when the air control system detects an alarm condition, the hot cells G.M. status is monitored in order to identify which one is currently running a radiosynthesis.

In the example described above the system will check the G.M. status in the selected hot cells (no. 3, 4, 7, 8). Let us suppose that the hot cell 3 alarm is ON, and a release of radioactive gas into the laboratory environment and/or the exhaust chimney is simultaneously detected: the ACS software will then immediately switch the corresponding hot cell three-way valve so as to deliver the contaminated air to the shielded gas cylinders. The ACS is based on two groups of four gas bottles with a volume of 50 l each (for a total volume of storable gas of 200 l for each group). Considering that the gas is compressed at the pressure of 200 bar, the storable volume of air pumped from the hot cell could be estimated at:

$$200\ \text{litres} \times 200\ \text{atmospheres} = 40{,}000\ \text{litres at atmospheric pressure}$$

Each hot cell has a total internal volume of approximately 300 l. With a pumping speed of 13,000 l h^{-1} and a pumping time of 20 min we obtain different air volume changes for the cells subordinately at the number of cells recognized to be in the synthesis phase.

When only one cell is in the synthesis phase, 15 air volumes are pumped out.

When two cells are in the synthesis phase, 7–8 air volumes from each cell are pumped out.

The counting volume for the contamination detection is a cylinder chamber of 2 l volume with an Na"(T") 2" × 2" detector with a Marinelli geometry, shielded by 5 cm lead over all the solid angle, with a minimum sensitivity at the energy of 511 KeV of 5 Bq/l. In practice, to keep the released contaminated air within the limit of 1 Bq/l, it is necessary to calculate a suitable threshold value that accounts for the dilution of the air in the pathway from the hot cells to the atmosphere. Considering that the outlet flow rate of the hot cell is 30 m^3/h and the flow rate of the general air conditioning line which arrives to the chimney is 55,000 m^3/h, we have a dilution factor of approximately 1/2000.

Therefore, an alarm threshold of 2,000 Bq/l at the hot cell will represent a suitable detectable signal to keep the radioactive concentration within the above limit.

We report and analyze here the results of one year of measurements; data obtained both before and after the installation of ACS will be reviewed.

The effective dose, due to intake and external irradiation, to the resident population included in an area within a radius of 100 m from the hospital chimney has been calculated.

3 Environmental Safety: Modelling of Dispersion in the Atmosphere of the Contaminated Gasses and Population Dose Estimation

The dose to the surrounding population from radioactive gasses released in the hot cells during synthesis has been calculated using the following model [3]:

- Conical dispersion with a diameter, at the measuring point, equal to 0.4 d, where d is the distance between the measurement point and the origin of the dispersion

- The air flows in a definite direction, according to statistics related to wind speed and direction typical of the considered area
- The entire activity is released in time t (approximately 1 h), equal to the population residential time

External irradiation:

$$D_\gamma = 0{,}0087 \cdot \frac{A_0 \cdot \Gamma}{d \cdot v}$$

Where:

A_0	Total exhausted activity	MBq
Γ	Gamma factor	mGy·m²/Mbq h
v	Wind speed	1 m/s
d	Distance from the origin	Meters

Intake:

$$D_{Int} = \frac{A \cdot R}{V} \cdot h(g)$$

Where:

Source activity	A_0
Conical dispersion volume at a distance d	$V = 0.0419\ d^3 (m^3)$
Wind speed	v = 1 m/s
Dispersion time flow	t = d/v(h)
Activity exhausted	$A = A_0 \cdot t$
Air volume inhaled at the time of the event duration (h)	$R = (1.2\ m^3/h)$
Dose factor for inhalation	$h(g)_{ina} = Sv/Bq$

Using this model, the dose to the population has been calculated for different release scenarios.

In our calculations the number of events is based on a statistical analysis of the justification file of our measurements in the period 1st January–30th June 2004 (no storage system installed) and 1st June–31st December 2005 (storage system installed and functioning).

Routine radioisotope production schedule is as follows (Table 1):

Table 1 Daily radioisotope production schedule

Production time	Isotope	Activity (Ci)
6.30	F18	15
7.45	F18	1.3
8.30	C11	1.0
12.00	F18	1.0
15.00	C11	1.0

Therefore we routinely perform two production per day of C-11 labelled radiopharmaceuticals and 3 production per day of F-18 derivatives, for an overall monthly estimate of 40 C-11 preparations, with a mean value of activity of 1 Ci per synthesis, and 60 F-18 preparations, with a mean value of activity of 1.3 Ci per synthesis.

Data analysis of the considered time frame, before the storage system installation, shows a 100% probability of a release of contaminated air (with a peak concentration > 500 Bq/l at the chimney) during the C-11 synthesis, while the probability of release of contamination is much lower during the synthesis of FDG, with a strong dependence on the quality of the commercial module selected. A mean value of 10/60 (16%) has been derived from the measurements file as the probability of a release of contaminants during the FDG synthesis.

The number of events per month represented in Tables 3 and 4 has been consequently assumed to be 40 for C-11 and 5 for F-18.

Obviously, the calculated dose values are proportional to the assumed number of events.

The Total Activity exhausted outside the chimney is simply calculated by the formula:

$$A_r = \left(C_m\left(Bq/1\right)\times F_a\left(1/h\right)\right)\times T$$

Where A_r represents the total activity released during an event with a duration T, with a measured concentration C_m, in a flow of exhausted air F_a.

The distance assumed in the calculation is the minimum distance between the top of the chimney and the nearest buildings surrounding the facility, about 150 m. The dose from external irradiation is inversely proportional to the distance, while the intake dose is proportional to the cube of the inverse of the distance. The outlet air flow of the air conditioning system servicing the controlled areas of the isotope production facility ranges between 3.3×10^7 and 5.5×10^7 l $\times$ h^{-1}. Therefore, if we measure 100 Bq/l for 1 h we should have a release of 5.5×10^9 Bq as A_r.

This formula has been used to calculate the values of the third and fourth columns of Tables 2 and 3. The lines indicating the maximum probability of our measurements are highlighted in Tables 2 and 3.

From the above data, it emerges that in order to keep the maximum effective dose to people living in the proximity of the facility below the limit of 10 μSv year^{-1}, a storage system to recover contaminated air released during the synthesis under ordinary working conditions becomes mandatory, especially due to the high contribution of the C-11 synthesis.

In fact, our estimation is a total annual dose to the neighbouring area in the range between 35 and 70 μSv year^{-1}.

After this evaluation it was decided to install such a system choosing among those commercially available. In October 2004 the system was installed and the data concerning the exhausted contaminated air, and consequently the dose to population, has been modified.

Table 2 C-11 release generated doses

Average act concenterate at chimney (Bq/l)	Release duration (min)	Total released activity (mCi)	Total released activity (Bq)	Number of events/month	External monthly dose (uSv)	Monthly intake dose (uSv)	Total yearly dose (uSv)
100	10	25	9.19E + 08	40	3.40E – 01	4.16E – 02	4.58E + 00
200	10	50	1.84E + 09	40	6.80E – 01	832E – 02	9.16E + 00
400	10	99	3.68E + 09	40	1.36E + 00	1.66E – 01	1.83E + 01
100	20	50	1.84E + 09	40	6.80E – 01	8.32E – 02	9.16E + 00
200	20	99	3.68E + 09	40	1.36E + 00	1.66E – 01	1.83E + 01
400	20	199	7.35E + 09	40	2.72E + 00	3.33E – 01	3.66E + 01
100	30	75	2.76E + 09	40	1.02E + 00	1.25E – 01	1.37E + 01
200	30	149	5.51E + 09	40	2.04E + 00	2.50E – 01	2.75E + 01
400	30	298	1.10E + 10	40	4.08E + 00	4.99E – 01	5.49E + 01

Table 3 F-18 release generated doses

Average act concenterate at chimney (Bq/l)	Release duration (min)	Total released activity (mCi)	Total released activity (Bq)	Number of events/month	External monthly dose (uSv)	Intake monthly dose (uSv)	Total yearly dose (uSv)
100	3	7	2.76E + 08	10	2.55E − 02	5.26E − 02	9.37E − 01
200	3	15	5.51E + 08	10	5.10E − 02	1.05E − 01	1.87E + 00
400	3	30	1.10E + 09	10	1.02E − 01	2.11E − 01	3.75E + 00
100	10	25	9.19E + 08	10	8.50E − 02	1.75E − 01	3.12E + 00
200	10	50	1.84E + 09	10	1.70E − 01	3.51E − 01	6.25E + 00
400	10	99	3.68E + 09	10	3.40E − 01	7.02E − 01	1.25E + 01
100	20	50	1.84E + 09	10	1.70E − 01	3.51E − 01	6.25E + 00
200	20	99	3.68E + 09	10	3.40E − 01	7,02E − 01	1.25E + 01
400	20	199	7.35E + 09	10	6.80E − 01	1.40E + 00	2.50E + 01
100	30	75	2.76E + 09	10	2.55E − 01	5.26E − 01	9.37E + 00
200	30	149	5.51E + 09	10	5.10E − 01	1.05E + 00	1.87E + 01
400	30	298	1.10E + 10	10	1.02E + 00	2.11E + 00	3.75E + 01

Table 4 Dose estimation from C11 compound emissions

Average act concenterate at chimney (Bq/l)	Release duration (min)	Total released activity (mCi)	Total released activity (Bq)	Number of events/ month	External monthly dose (uSv)	Monthly intake dose (uSv)	Total yearly dose (uSv)
200	10	50	1.84E + 09	20	0.339919	0.0415843	4.57

Table 5 Dose estimation from F18 compound emissions

Average act concenterate at chimney (Bq/l)	Release duration (min)	Total released activity (mCi)	Total released activity (Bq)	Number of events/ month	External monthly dose (uSv)	Monthly intake dose (uSv)	Total yearly dose (uSv)
100	5	12	4.59E + 08	15	6.37E − 02	1.32E − 01	2.34E + 00

The installation of an automated Control and Storage System has greatly decreased both the number and the characteristics of the release events; that is peak value and duration.

For C-11 complete control was obtained (no peaks at all at the chimney) in 50% of the syntheses with a decrease of the mean duration time to 10 min.

For F-18 a decrease of the event duration to 5 min was obtained, with a reduction of the maximum to a mean value of 100 Bq/l.

Consequently, the figure for the dose to population is now expressed by the values reported in Tables 4 and 5.

4 Safety and Shielding of the Hot Cells

Standard cell for synthesis has a thickness of 75 mm lead. This thickness achieves a maximum external rate below 10 μSv/h, when stored activities are of the order of 10 Ci (370 GBq).

The safety of the modern hot cell is software driven. The main safety features are:

- Front window opening not allowed when
  - Radioactivity transfer is in progress
  - Internal dose rate exceeds threshold
- The transfer of radioactivity from the cyclotron is allowed only when the front window is closed and the cell is fully sealed.

5 The Staff Internal Contamination Risk During Synthesis

In case of detected leakage of contaminated air, personnel attending the synthesis in the Radiochemistry Lab are requested to take a urine sample at 1 h after the accident. The urine samples are measured by means of a Na-I gamma counter (LKB Wallac energy resolution = 8.4 keV and efficiency = 9% at 661 keV of ^{137}Cs). The measurements are performed on 2–5 ml of urine with an energy window covering the whole spectrum and acquisition time of 5 min. With this set up the efficiency of gamma counter is 82.3% with an MDA of 5 Bq/cc.

The staff of the radiochemistry labs has been monitored over the last 3 years also with WBC (Joint Research Centre ISPRA, Italy; MDA for the energy of Cs-137, F-18 and C-11 of 20 Bq Body counting and an estimated error of 5%). The goal of these measurements was to monitor the risk of intake of inhaled contaminated powders occurring during cyclotron maintenance, and also to discover low level contamination from routine isotopes production.

In addition, to establish the percentage of urine excretion in case of inhalation of F18 where no specific references have been found, a biodistribution study in nude mice was performed. Twelve nude mice were placed in a hot cell and air contamination with ^{18}F was produced. The animals (in three groups) were sacrificed at 10, 30 and 60 min from the beginning of air contamination. Activity in the kidneys, lungs, bladder, urine, bone, muscle and blood was determined by a gamma counter. For blood, bone and muscle the measurement was performed on samples and the activity was normalized to the organ weight [2]. The whole body activity for each animal was calculated as the sum of activity in all considered organs. The activity in each organ was normalized to the whole body activity.

6 Modelling for Dose Determination from Urine Sample Measurement

6.1 F-18 Model

F-18 is considered to be present in the contaminated air mainly in the form of very small droplets of aqueous HF generated inside the target during the cyclotron proton irradiation. Analyzing data obtained by an air monitoring system, two peaks corresponding to released activity from the radiosynthesis automated modules are in fact typically detected inside the hot cell. The timing of the first peak matches with the F-18 activity transfer from the cyclotron target to the hot cell, while the second peak corresponds to the solvent removal by evaporation step.

The modelling of the kinetic takes into consideration the following [4]:

- Comp 1: Extra-thoracic Airways + Lung
- Comp 2: Blood
- Comp 3: Urine

As described by [5], the respiratory tract uptake, when particle size has an AMAD of 5 μm and is deposited according to the model, produces different deposited fractions in the anatomical regions defined by the ICRP 66 model with a total deposition fraction of 0.82. This fraction will be considered in the calculation of intake from air contamination values leading to the formula concerning the estimation of committed effective dose here reported:

$$\text{Eff. Dose}\,(\text{Sv}) =\text{:in haled activity}\,(\text{Bq}) \times 0.82 \times 5.4 \times 10^{-11}\,\text{Sv} \times \text{Bq}^{-1}$$

where the last value represents the committed effective dose per unit of F-18 intake considering rapid absorption [6].

Data on urine excretion after F-18 inhalation are not reported in literature. Potter [7] demonstrated that the whole body intake retention fraction (IRF) for non radioactive Fluorine (class F) is equal to 0.481 at 6 h after inhalation, and remains constant for a long period, while the rest is eliminated through the lungs and airways with a long half-life compared with the physical half-life of F-18. No excretion values for urine and feces are given. Charkes [8] stated that the percentage of cumulative urine excretion in case of intravenous administration of F-18 is 11% at 1 h post-injection. The results obtained from our animal studies have been weighted in terms of percentage of intaken activity as follows:

$$\text{Organ activity} / (\text{inhaled activity mice} \times 0.82)$$

6.2 *C-11 Model*

C-11 is, in most cases, produced "in-target" in the chemical form of [^{11}C]CO_2 by proton irradiation of a gas mixture containing N-14 and small percentages (1–2%) of oxygen. Alternatively, it may be obtained in the form of [^{11}C]CH_4 when oxygen is replaced by hydrogen (5–10%). Both the targets installed at San Raffaele Hospital yield [^{11}C]CO_2, which is rapidly transferred at the end of bombardment to the radiosynthesis module located in the hot cell. The most critical point from the air contamination point of view is the arrival of [^{11}C]CO_2 and its trapping with liquid nitrogen or its reaction with hydrogen to form [^{11}C]CH_4, depending on the chosen radiosynthetic pathway. The major contaminant is thus [^{11}C]CO_2 itself. Carbon dioxide is known to be highly soluble in water, and its physical form can be freely diffusible gas and/or dissolved in very small water droplets. The average relative humidity of the radiochemistry lab is typically 50%, while the air exchange rate is about 20 air exchange/hour. Whatever its physical form, CO_2 is rapidly distributed throughout the body when inhaled. The concentration of carbon dioxide in the plasma is three times greater than that in red blood cells. The gas is carried partly in solution (2.4–2.7 vol %), but mostly either as bicarbonate (42.9–46.7 vol %), or as carbamino compound (3.0–3.7 vol %).

The dose to organ could be considered as proportional to the total activity trapped by the organ as defined by the following formula [9, 11]:

$$\mathrm{D} = \overset{*}{\mathrm{A}}\mathrm{L} \times \frac{c}{M_t} \sum_i E_i Y_i \phi_i \tag{1}$$

Where

$$\overset{*}{A} = \int_0^t A(u)du \tag{1a}$$

represents the total number of decays in a time t from 0 to infinity, during which the radionuclide is inside the organ,

c is a constant = 1 for quantities expressed in the SI unit
M_t is the mass of the target organ
E_i is the mean energy of the radiation of type I
Y_i is the yield of the radiation type i
ϕ_i is the absorbed fraction of energy of type i radiation

When dealing, as in our case, with a single type of radiation, formula (1) can be reduced to:

$$\mathrm{D} = \overset{*}{\mathrm{A}}\mathrm{L} \times \frac{c}{M_t} EY\phi$$

We shall consider kidneys and lungs as target organs. The masses of the two organs in the average man are respectively 300 and 1,000 g.

The product $EY\phi$ is given by the term Y, which is constant, and the product $E\phi$, which represents the fraction of the emitted energy released in the organ. We can introduce a factor K defined as:

$$K = \frac{\rho_{lung} \times r_{lung}}{\rho_{kidney} \times r_{kidney}} = \frac{E\phi_{lung}}{E\phi_{kidney}} = \frac{0,25 \times 6}{1,1 \times 3} = 0,455$$

In fact, the third term of this expression represents the fraction of energy released in the organ for each disintegration. Considering that at a photon energy of 511 KeV, corresponding to the quanta energy generated by the positron annihilation, the dominant effect is the Compton effect, the ratio represented by the third term can be equated to the second term of the expression.

For lungs and kidneys, therefore, we can state:

$$\frac{Dose_{lung} \times M_{lung}}{Dose_{kidney} \times M_{kidney}} = \frac{\overset{*}{A}_{lung}}{\overset{*}{A}_{kidney}} \times K \tag{2}$$

The ICRP 53 [4] gives the time of permanence of the total activity inside the organ after inhalation: for continuous breathing of the gas for 1 h the respective times are:

Lung	12.5 s
Body (assumed for kidney)	9.3 min
The ratio between the two times is	45

Bearing in mind that

$$\overset{*}{A} = \int_0^t A(u)du$$

it can also be expressed as

$$\overset{*}{A} = \int_0^t A(u)du = \mathrm{A}_{\mathrm{mean}} \times \text{time of permanence in the organ}$$

Equation (2) becomes:

$$\frac{Dose_{lung} \times M_{lung}}{Dose_{kidney} \times M_{kidney}} = \frac{A_{meanlung}}{A_{meankidney}} \times \frac{timelung}{timekidney} \times \mathrm{K} \tag{3}$$

$$\frac{Dose_{lung} \times M_{lung}}{Dose_{kidney} \times M_{kidney}} \times \frac{timekidney}{timelung} \times \frac{1}{K} = \frac{A_{meanlung}}{A_{meankidney}} \tag{4}$$

In conclusion, the ratio of the organ doses times the ratio of the respective masses, times the inverse of time ratio, times 1/K, is approximately referable to the ratio of the activity trapped by the two organs.

Furthermore, we have added the hypothesis that all the activity concentrated in the kidneys will be excreted with urine. Therefore the ratio of formula (2) indicates the ratio between the lung activity content and the cumulative urine excretion.

From the data of ICRP 53 and 80 concerning the inhalation of $C^{11}O_2$, considering the masses of the two organs (1,000 g for the lungs and 300 g for kidneys) and the organ permanence time, the first term of Eq. (4) is equal to approximately 96.5, which we shall consider 100.

Therefore we can assume that the urine C-11 content will represent 1.0% of the fraction of the inhaled quantity deposited in the lung (intake), in agreement with the data published by Legget [10], who estimates an expected excreted activity amount ranging between 0.5% and 3.2% between 6 and 24 h after intravenous administration in mice.

Considering the short decay of C11 ($T_{½}$ = 20.38 min) the measurement of these activities must be performed a maximum of 1 h after the inhalation. This implies a reduction of the expected amount of the excreted activity. However, even considering the abovementioned reference, the initial value of 1% at 1 h seems sufficiently conservative.

6.3 *A Heuristic[1] Model for F-18 and C-11, Which Considers the Following Hypotheses*

- The test is performed within 1 h after the inhalation event.
- At 1 h a cumulative excretion of 5.0% of the absorbed (retained from body) activity for F18 and 1.0% of the absorbed (retained from body) activity for C11, both corrected for respective decay times.

Under these hypotheses, assuming an intake corresponding to a committed effective dose of 1 mSv, taken from the data reported on ICRP 68 concerning the intaken activity to dose coefficients (Sv/Bq) (in our formula referred to as Hg), reference values for urine concentration at different times after inhalation have been calculated using the formula both for F-18 and C-11:

- $\mathbf{(Hg)^{-1}}$ (Bq/Sv) × $\mathbf{10^{-3}}$ (Sv/mSv) × $\mathbf{a_s}$ × (l^{-1}) =: **Bq** corresponding to effective Commulative Dose of 1 mSv

  Where:

  Hg is a coefficient representing the committed effective dose (in Sv) for inhaled Bq. It results that $(Hg)^{-1}$ will represent the amount of inhaled bequerel which will give a committed effective dose of 1 Sv.

  $\mathbf{10^{-3}}$ is the factor converting the effective committed dose to 1 mSv.

  $\mathbf{a_s}$ is the considered amount for cumulative excretion at time t after inhalation.

The resulting data is presented in Table 6.

The two methods, ICRP derived Heuristic and Environmental, will be tested analyzing positive contamination results obtained last year after gaseous dispersion in the lab.

An estimation of the committed effective dose due to micro dispersion events calculated under the hypothesis described here is also presented in order to obtain a cautious estimation of the committed effective dose to the radiochemists in the absence of positive test results.

Table 6 Urine activity content

Isotope	Urine activity content at 1 h (Bq)
C-11	4.07 E05
F-18	6.34 E05

[1] Note: **heuristic** refers to a method, commonly informal, to help solve a problem. It is particularly used for a method that often rapidly leads to a solution that is usually reasonably close to the best possible answer. Heuristics are "rules of thumb", educated guesses, intuitive judgments or simply common sense.

This estimation has been obtained considering the following hypotheses for the calculations:

- The average time considered to be spent by the radiochemist inside the lab during synthesis is 60 min per day
- The inhaled volume is 600 l/h (1/2 of the value used for dispersions above threshold).
- The occurrence of micro-contamination events for each radiochemist is estimated at 200 days/year.
- The retained fraction is estimated to be 50% of the inhaled amount with fast absorption rate [11, 12].

These data have been utilized to validate the model of chronic exposure.

7 Results and Discussion

In the last 18 months we have experienced four (three with F-18 and 1 with C-11) events of leakage of contaminated air inside the Radiochemistry lab during synthesis with coincident presence of personnel inside, with around 1,000 synthesis processes during the year.

These data confirm that when the hot cells are well controlled and regularly checked by means of thorough preventive maintenance and the synthesis modules are well designed, the risk of releasing contaminating activity into the radiochemistry lab is quite low, in our case of the order of 1E-02

However, it is impossible to entirely eliminate the risk of heavy dispersion in the labs with an activity of some 500 GBq/day being synthesized, and amounts up to 5% of the synthesized activity possibly flowing from the module both during the activity transfer from the cyclotron and during the synthesis process.

7.1 F-18

In Table 7 the activity, expressed as percentage of whole body intaken activity, in the considered organs at different time obtained in nude mice after inhalation of F-18 is shown. At 10 min the maximum F-18 uptake was found in the bone

Table 7 Organ data in mice

	(%) of absorbed activity by inhalation ± sd		
Organ	10 min	30 min	60 min
Blood	8.86 ± 2.86	5.37 ± 1.51	0.80 ± 0.22
Lung	1.37 ± 1.22	0.25 ± 0.11	0.13 ± 0.05
Urine	5.19 ± 0.81	7.27 ± 2.21	5.89 ± 3.51
Kidney	2.49 ± 0.66	0.76 ± 0.34	0.32 ± 0.28
Bone	65.08 ± 4.05	68.19 ± 3.77	72.62 ± 5.18

Table 8 F18 contaminations

Episode #	Average air concentration (Bq/l)	Presence of the staff (min)	Estimated dose from lab data (uSv)	Urine content (Bq)	Estimated dose from urine μSv Heur. meth. / ICRP data	Operator #
1	480	5	2.6	1,318	2.08	1
			2.6	1,124	1.77	2
			2.6	1,884	2.97	3
2	1,340	5	7.51	1,600	2.52	1
3	1,390	5	7.24	2,627	4.14	1
	1,390	5	7.24	350	0.55	2
	1,390	5	7.24	805	1.27	3

(approaching 70%) and remained constant until 90 min. These data suggest a very fast clearance of the F-18 from the lung to the blood with a consequent uptake to the bone. This behaviour reflects the kinetics of F-18 when injected directly into the blood compartment [4]. The ratio of the residual lung activity to blood content, at the maxima (after 10 min) ranges between 1:10–1:8 while the urine excretion at 60 min derived from these data is 5.89% ± 3.51%.

The rounded mean value of 5% is used to evaluate the dose from urine samples at 1 h for our calculations. In Table 8 a comparison between estimated doses from measured air contamination environmental levels and the Heuristic model calculated results is reported.

The wide variations among the urine data for the third event are likely correlated to a difference in the amount of inhaled radioactivity among the three operators. This difference cannot be considered when the doses are evaluated from the dispersion data. In fact in this case the duration of the event is assumed to be equal to the total time of presence of the staff inside the lab.

7.2 C11

The same methodology used for F18 intake dose evaluations and comparisons between different models leads to the results reported in Table 9.

7.3 Chronic Low Level Exposures Data

As a further source of experimental data we used the data coming from the WBC carried out at ISPRA. During these measurements, designed to check the presence of long lived isotopes eventually inhaled during target maintenance, five positive cases were registered over 17 measurements; in three cases the contamination was detected in the lungs and in one case in body (lung + kidneys + bladder). These data are presented in Table 10.

Table 9 C11 contaminations

Episode #	Average air concentration (Bq/l)	Presence of the staff (min)	Estimated dose from lab data (uSv)	Urine content (Bq)	Estimated dose from urine μSv Heur. meth. / ICRP data	Operator #
1	11507	5	1.67	1201.72	5.89	1
			1.67	999.36	4.90	2

Table 10 Isotopes inhaled during target maintenance

Date	Detected activity (Bq)	Site	Isotope
13/09/2005	7	Lung	F-18
15/12/2005	7	Lung	F-18
15/12/2005	6	Lung	F-18
16/03/2006	525	Hand	F-18
16/03/2006	40	Body	F-18

Table 11 Effective committed dose from low level chronic exposure

Isotope	Average environmental concentration (BqL^{-1})	Inhaled activity (Bq)	Coefficient intake dose ($Sv\ Bq^{-1}$)	Dose ($\mu Sv\ year^{-1}$)
C11	10	6.00E + 05	3.20E − 12	1.92E + 00
	50	3.00E + 06	3.20E − 12	9.60E + 00
	100	6.00E + 06	3.20E − 12	1.92E + 01
F18	10	6.00E + 05	5.40E − 11	3.24E + 01
	50	3.00E + 06	5.40E − 11	1.62E + 02
	100	6.00E + 06	5.40E − 11	3.24E + 02

These data led us to define a provisional reference table for effective committed dose from low level chronic exposure, where the intaken activity was calculated for different air concentrations of the contaminants C-11 and F-18 (see Table 11), and under the assumptions defined in the previous paragraph for chronic low level exposure [12–14].

The probability of dispersion events above the alarm threshold value (100 Bq/l) is of the order of 1/100, considering that we experienced four events over 600 syntheses during 2007, while the occurrence of micro dispersion below the threshold limit (100 Bq/l) in the synthesis labs has a probability close to 100%, as demonstrated by the data collected at our facility.

When considering the high possibility of the occurrence of these events, a cautionary estimated dose for staff involved in the syntheses, for high workload installations, must be calculated in the range between 100 and 500 μSv/year [14].

8 Conclusions

In our experience it has been demonstrated that even very simple models can provide good inter-correlation in the evaluation of the effective committed dose.

In fact, at least for radioprotection purposes, and in the range of small dispersions where the models have been tested, the agreement among data seems acceptable even when compared with data obtained from the radioactivity concentration and effective presence of the personnel in the labs.

In our cases, when considering the environmental data to calculate the intake of F18, these always represent an overestimation of the committed effective dose when compared with the calculated values from urine samples, even if different models are used.

However, an opposite behaviour has been observed in the data of committed effective dose from C11 inhalation. The greater weakness of the C11 model is marked also by the lack of experimental data derived from intake measures.

It should be possible in future studies to verify the assumption of 1.0% of the inhaled activity as the cumulative urine excretion 1 h after inhalation. A higher value, up to 2.5%, could be reasonable, leading to a better approximation of the committed dose calculations derived from the environmental data, but such a value does not currently seem derivable from the bibliographic references considered.

As F18 is concerned, we can argue that the data deriving from environmental dispersion values can be assumed as worst cases for operator doses, whereas the testing of urine samples, when available, must be considered more realistic and appropriate for accurate individual dose evaluation.

In fact, the effective time of operator presence, along with breathing modality, can introduce great uncertainties into the evaluation derived from air contamination data, whereas urine samples represent robust data from which to calculate the inhaled radioactivity amount, and from these, the committed effective dose, although an approximation factor of 2 currently seems reasonable; in fact more data from urine samples of contaminated radiochemists will be necessary to definitively validate our assumptions of 5% for F18, and 1% for C11 as excretion percentages.

Whenever urine concentration data are unavailable, committed effective doses can be calculated from data dispersion, but in this case continuous and accurate air monitoring in the lab must be installed and calibrated.

Finally, we would underline the criterion of leaving the lab whenever possible during the synthesis process in order to limit low chronic intake and related additional committed dose to the staff.

References

1. Calandrino R., del Vecchio A., Todde S., Fazio F.; Measurement and control of the air contamination generated in a medical cyclotron facility for PET radiopharmaceuticals. Health Phys, 2007 May; 92(5 suppl 2): S70–S77.
2. Calandrino R., del Vecchio A., Savi A., Todde S., Belloli S.; Intake risk and dose evaluation methods for workers in radiochemistry labs of a medical cyclotron facility. Health Phys, 2009 (in press).

3. Gazzetta Ufficiale della Repubblica Italiana, D. Lgsl 230/95, 17th march 1995. Birattari C., Bonardi M., Ferrari A., Silari M.; Neutron activation of air by a biomedical cyclotron and assessment of dose to neighbourhood populations. Radiat Protect Dosim, 1986; 4: 311–319.
4. IAEA Safety Reports Series No. 37 Methods for Assessing Occupational Radiation doses due to Intakes of Radionuclides (2004).
5. ICRP Human Respiratory tract Model for Radiological Protection, Publication 66, Pergamon Press, Oxford (1994).
6. ICRP, Dose Coefficients for intakes of radionuclides by workers. Publication 68, Pergamon Press, Oxford (1994).
7. Potter et al.; Intake retention fractions developed from models used in the determination of dose coefficients developed for ICRP publication 68- Particulate inhalation Health Phys, 2002; 83(5): 594–789.
8. Charkes et al.; Studies of skeletal tracer kinetics.1: Digital-Computer Solution of a Five compartment model of F-18 fluoride kinetics in humans. J Nucl Med, 1978; 19: 1301–1309.
9. ICRP, Radiation Dose to Patients from Radiopharmaceuticals. Publication 53, Pergamon Press, Oxford (1987).
10. Leggett R.W.; A biokinetic model for Carbon Dioxide and Bicarbonate. Radiat Protect Dosim, 2004; 108(3): 203–213.
11. ICRP, Radiation Dose to Patients from Radiopharmaceuticals. Publication 80, addendum 5 and 6 to Publication 53. Pergamon Press, Oxford (2001).
12. ICRP, Limits for intakes of radionuclides by workers. Publication 30, Pergamon Press, Oxford 1979.
13. ICRP, Individual Monitoring for intakes of radionuclides by workers: design and interpretation. Publication 54, Pergamon Press, Oxford and New York (1988).
14. Lopez Ponte M.A. et al.; Individual Monitoring for internal exposure in Europe and the integration of dosimetric data. Radiat Protect Dosim, 2004; 112: 69–119.

Optimisation of Patient and Staff Exposure in Interventional Radiology

Renato Padovani

1 Introduction

The number of fluoroscopically guided interventional cardiology procedures increased more and more rapidly in the last 10 years together with their complexity. The main reason is that, with interventional cardiology, even more patients can often be cured without the use of surgery and their stay in hospital is limited.

In complex procedures patient and staff can receive high radiation dose if proper quality assurance and training programmes are not in place.

The radiation dose to patient depends on a number of factors, including patient size, equipment, technique and type of examination. Large variation in patient dose, for the same type of X-ray examination, have been demonstrated in several studies [1–5]. These variations are almost due to different complexity of the procedures, equipment performance, procedure protocols and patient body size.

By investigating patient dose, variations can be acknowledged, causes founded and the necessary adjustments can be implemented.

Reference Levels (RL) provide a framework to reduce this variability and assist in the optimisation process [6–8]. For this reason, monitoring patient exposure in prolonged interventional procedures and comparison with RLs is a mandatory task in every quality assurance programme.

The staff operates near the patient and is exposed to a non uniform radiation field due to patient scattered radiation. Consequently workers may receive, over a period, relatively high radiation doses [8–10].

According to the European Union Directive 96/29/EURATOM, radiation dose to workers has to be expressed in term of effective dose E (ICRP publication 60), quantity that is related to the stochastic radiation risk. The avoidance of deterministic effects is instead ensured through limits on equivalent doses H to few specific tissues (extremities, eye lens, skin).

R. Padovani (✉)
Medical Physics Department, University Hospital S. Maria della Misericordia, Udine, Italy
e-mail: padovani.renato@aoud.sanita.fvg.it

Y. Lemoigne and A. Caner (eds.), *Radiation Protection in Medical Physics*,
NATO Science for Peace and Security Series B: Physics and Biophysics 1,
DOI 10.1007/978-94-007-0247-9_20,

Since effective dose, weighted sum of several organ doses, and equivalent dose cannot be measured in practice, two operational quantities, personal dose equivalents $H_p(10)$ and $H_p(0.07)$, have been defined and recommended by the International Commission on Radiation Units and Measurements (ICRU) and ICRP [11] as conservative estimates of E and H_{skin}. Thus, effective dose and equivalent dose can in theory be known through readings from dosemeters appropriately located on the body.

Both E and $H_p(d)$ vary in a complicated way with radiation type and quality, energy spectrum, fluency rate and x-ray beam direction of incidence. In addition, the use of protective clothing and protective devices makes radiation exposure highly inhomogeneous and assessment of individual dose can result to a very complex task.

These special situations may require more than one dosemeter. Recommendations [11, 12] suggest that if a single dosemeter is used it should be worn outside the apron to monitor doses to skin, eye and unshielded parts of the body even if it will overestimate E. When overestimations are unacceptable two dosemeter should be used one over and one under the protective apron, ICRP Publication 75 [11]. The interpretation of combined results will have to depend on local irradiation condition.

Consequently several methods of measurements and algorithms have been proposed to obtain reasonable estimates of the effective dose [12–20] and a great variety of methods for assessing and type of monitoring programs has been identified among European countries and centres.

The lecture will highlight the followign topics:

1. Interventional radiology equipment performance assessement
2. Patient dose monitoring and reference levels
3. Staff dose monitoring and investigation and constaint dose values

2 IR Equipment Performance Assessment

The lecture will describe a recent survey performed by SENTINEL, a European research group.

A questionnaire was sent to a sample of European centres to collect dosimetry data (typical entrance air kerma rate in fluoroscopy and imaging mode), image quality evaluations (low and high contrast resolution) and KAP calibration factors. The questionnaire included instructions on the agreed methodology to be followed for measurements.

The list of angiographic units included in the survey is reported in Table 1 and comprises six systems with Flat Panel imaging detectors (FPD) and six with Image Intensifier-TV chains (II). The table reports also the year of installation.

Tests included measurement of air kerma dose rates in fluoroscopy and digital acquisition modes and a subjective assessment of image quality using the Leeds test object TOR 18FG. Dose rates were measured under Automatic Exposure Control

Table 1 Cardiac angiographic systems included in the SENTINEL survey

Unit no.	Manufacturer	Model	Imaging detector	Year of installation
1.	Siemens	Axiom Artis dBc	FPD	2005
2.	Siemens	Axiom Artis dBc	FPD	2005
3.	Siemens	Bicor Top	II	1995
4.	Siemens	Multistar T.O.P.	II	1995
5.	Philips	Allura F9	FPD	2002
6.	Philips	Allura 9	II	2002
7.	Philips	Integris 5000H	II	1998
8.	GE	Innova 2000	FPD	2002
9.	Philips	Integris 3000	II	1994
10.	Siemens	Axiom Artis	FPD	2003
11.	Philips	Integris CV9	II	2003
12.	Siemens	Axiom Artis	FPD	2004
13.	Philips	Integris 5000H	II	2002

(AEC) in fluoroscopy and digital acquisition modes by measuring the entrance surface air kerma rate when a phantom of 20 cm PMMA thickness simulates a patient attenuation and the field of view (FOV) on the detector has been set at 22 cm or nearest with a focus-entrance phantom distance of about 65 cm and the image detector positioned at 5 cm from the exit phantom surface.

With the purpose to use the KAP meter calibration factor to correct collected patient KAP values, the calibration procedure is performed taking into account the attenuation determined by the patient table and mattress. The calibration has been performed at 60–80–100 kV x-ray qualities with an ion chamber on the axis of the x-ray beam placed at minimum 10 cm away from the patient table and the image detector to avoid scatter. The different x-ray qualities are reached inserting in the x-ray beam, between the ion chamber and the image detector, attenuating material (copper and/or aluminium) simulating the patient attenuation and driving both kV and added filtration to typical clinical conditions. Surface area is calculated from field dimensions measured with a radio-opaque ruler or an equivalent method. KAP calibration factor is assumed as the mean value of the calibration factor measured for the 3 x-ray qualities.

2.1 Entrance Surface Air Kerma Rates

The majority of the tested systems has a wide range of user selectable dose options including a range of pulsed fluoroscopy modes, digital acquisition frame rates and automatic insertion of spectral filters. The pulsed fluoroscopy mode most frequently used on the equipment tested is 12.5 or 15 pulses per second (pps) and the acquisition modality 12.5 or 15 images per second.

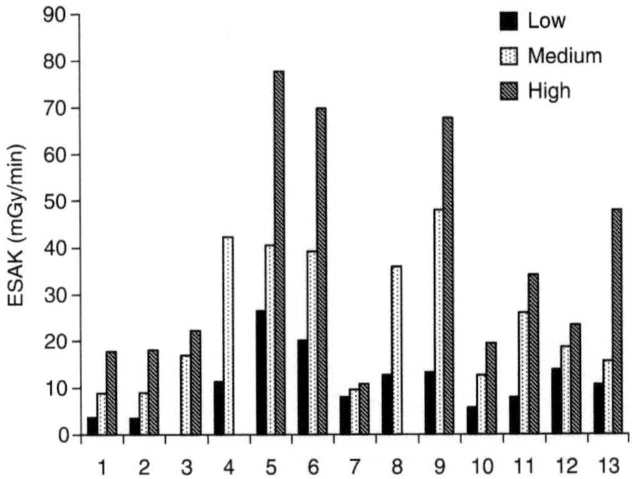

Fig. 1 Entrance surface air kerma rate in fluoroscopy for 13 cardiac angiographic systems at the entrance surface of a phantom of 20 cm of PMMA, FOV about 22 cm

Figure 1 reports entrance surface air kerma rate for different fluoroscopy modes available in each system. The air kerma entrance rates range from 3.6 to 26.5 mGy/min in low fluoroscopy mode, from 8.8 to 48 mGy/min in medium fluoroscopy mode and from 10.7 to 77.7 in high fluoroscopy mode. Air kerma entrance rate does not seem to be strictly manufacturer dependent. For the majority of the systems tested, the patient entrance dose rate varies between 5 and 20 mGy/min for low and medium modes. The two systems presenting the highest dose rates are installed in the same center.

The entrance surface air kerma per image was in the range 32.9–192 μGy/frame in low cine mode and 77.8–316 μGy/frame in normal acquisition mode.

2.2 Image Quality

Image Quality was assessed by imaging the Leeds test objects TOR 18FG. For all of the systems the threshold contrast varies between 2.5% and 4%. Only unit no. 3 has a threshold contrast quite lower (2.3%).

In general, an improvement in image quality is not apparent for the systems operating at higher dose level. This is particularly important for systems exhibiting the highest entrance doses. All analysed systems have limiting spatial resolution greater than 1.25 lp/mm.

Table 2 Reference levels proposed for interventional cardiology equipment

Imaging mode	Entrance surface air kerma rate (ESAK)
Fluoroscopy low	13 mGy/min
Image acquisition	100 μGy/frame

2.3 Kerma Area Product Meter Calibration

A large variation, $KAP_{real}/KAP_{dispalyed}$ from 0.68 to 1.05, in KAP meter calibration and/or in the attenuation properties of patient tables and mattresses is recognised and cannot be neglected when patient doses are reported or compared between centres.

The survey on the cardiac angiographic units in a sample of European centres demonstrates a large variability in entrance dose rates for both, fluoroscopy and image acquisition modes, image quality performance and KAP calibration.

As an outcome of this study a preliminary set of reference levels for the Entrance Surface Air Kerma (ESAK) quantity is proposed in Table 2; it can be adopted by centres and maintenance engineers to set up cardiac equipment at an acceptable dose performance level and by standardisation bodies as an input to introduce proper standards.

3 Patient Exposure and Reference Levels in Interventional Cardiology

Patient dosimetry in interventional radiology is performed with the following aims:

- Quality assurance: to compare patient doses with reference values or with other centres or other type of procedures
  - Dose quantities: air kerma area product (KAP) and the cumulative dose (CD) at the interventional reference point (IRP); dose analogous: fluoroscopy time, number of images, number of series
- Stochastic risk evaluation: to assess organ doses and effective dose
  - Methods: phantom measurements or/and monte carlo simulation dosimetry methods
- To prevent determinist injuries to skin
  - Dosimetry methods: CD at IRP and area dosimetry with large radiochromic films

As an example, the experience of a survey performed in Europe to assess reference levels in interventional cardiology in the following procedures is reported:

- Coronary angiography (CA)
- Percutaneus transluminal coronary angioplasty (PTCA)

- Electrophysiology procedures, including diagnostic electrophysiology, pacemaker implantation (PM), defibrillator implantation (ICI) and radiofrequency cardiac ablation (RFCA)

The survey involved nine European partners and near 2,000 procedures were examined. Information, including the fluoroscopy time, number of frames, air kerma-area product (KAP), and, when available, the cumulative dose (CD) to Interventional Reference Point (IRP), were provided. Accuracy of dose values provided have been submitted to a dosimetry intercomparison.

3.1 Coronary Angiography and PTCA Procedures

Examined dose or dose analogues data exhibit a large variability. In Figs. 2 and 3 mean and median values of fluoroscopy time (FT) and KAP, respectively, are reported for CA procedures.

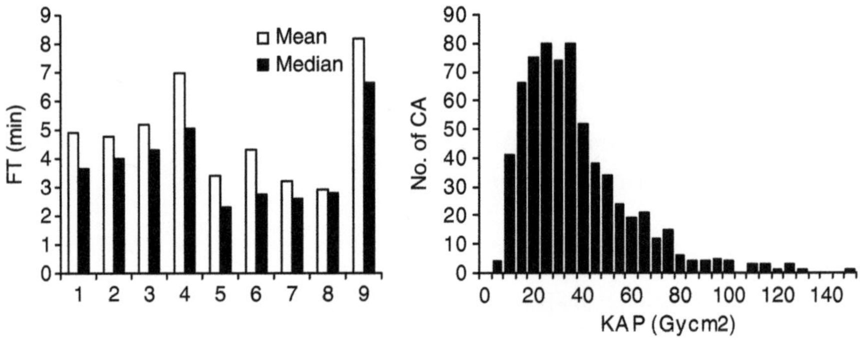

Fig. 2 Mean and median fluoroscopy time and air kerma-area product of CA distribution in the participating centres

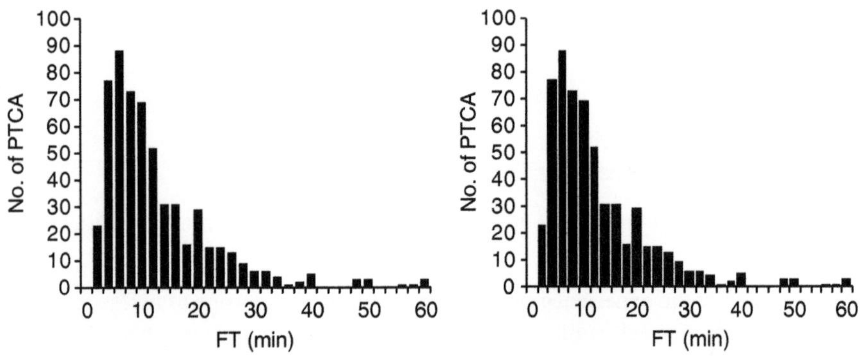

Fig. 3 Frequency distribution of fluoroscopy time and KAP for PTCA procedures in participating cardiac centres

Table 3 SENTINEL reference levels for interventional cardiac procedures

	Procedures		
Dose or dose analogue	CA	PTCA	EFO
KAP ($Gycm^2$)	45	85	35
Effective dose (mSv)	8	15	6
CD at IRP (mGy)	650	1,500	–
Fluoroscopy time (min)	6.5	15.5	21
Number of cine images	700	1,000	–
Entrance surface air kerma rate	Fluoro low: 13 mGy/min Image acquisition: 0.10 mGy/fr		

The examinations have been pooled and frequency distribution of fluoroscopy time, number of frames and KAP derived together with the associated reference levels.

3.2 Reference Levels

In Table 3 reference levels, assessed as the rounded value of the 75th percentile of the distributions, are reported for fluoroscopy time, air kerma-are product, cumulative dose at IRP, and No. of cine images.

Because equipment performance and equipment set up by the maintenance service is one of the factors contributing to patient dose variability, entrance surface air kerma for fluoroscopy and image acquisition, measured at the entrance of a 20 cm PMMA phantom, are also introduced in the set of proposed reference levels.

The set of reference levels proposed for coronary angiography and angioplasty are lower compared to those assessed in 2004 by the DIMOND group (CA: KAP = 57; PTCA: KAP = 94 $Gycm^2$) [7]. The main difference derives from the lower number of cine images that had influenced the KAP.

Regarding the introduction of the cumulative dose at IRP in the set of reference levels, it is necessary to better evaluate the impact of this quantity in the optimisation process of patient exposure.

4 Staff Exposure, Intervention and Constraint Dose Values

Staff exposure in interventional radiology is an important and actual topic in radiation protection. Recently several studies have been undertaken with the aim to evaluate the status of staff monitoring, to assess accuracy of dosimetry methods and to propose acceptable levels of optimised exposure.

As an example, the lecture will describe the methods and the results of a Europen survey performed to investigate the methods for measuring doses received by the staff employed in interventional cardiac laboratories and the algorithms used for estimating E. For this purpose a questionnaire has been sent in 2005 to more than 20 centres participating in the SENTINEL project. The questionnaire asked for general and technical information about type of personal and area dosimeters, dose quantity measured, dose calculation methods and monitoring programme details.

A second questionnaire was intended to collect staff exposure data. Radiation dose measurements performed in 12 European country have been collected. Data have been gathered over a period of at least 2 years for 2–4 cardiologists in each centre.

E is calculated using the Niklason algorithm [14]when two measurements, one over apron-collar (H_o) and one under waist (H_u), are available:

$$E = 0.02(H_o - H_u) + H_u$$

If only one dosemeter is used under the apron, the effective dose is estimated as $H_u/21$, as recommended in NCRP report 122 [12].

4.1 Survey on Methods for Assessing the Staff Doses due to External Exposures in Interventional Cardiology Procedures

Thermoluminescent dosimetry (TLD) is the most widely used technology for personal dosimetry. In all investigated centres, TL dosimetry is adopted and the TL material mainly used is LiF:Mg.

In spite of this uniformity in dosimetry technology adopted, staff dosimetry is performed with different modalities, as emerged from information collected and reported in Table 4.

Only in seven of the investigated centres two personal dosimeters, one under and the other over the apron, as recommended by ICRP, are used for the estimate of the effective dose. In five centres a single dosimeter is used for the assessment of effective dose: in two of these centres the dosemeter is worn over the protective apron and in the other three is worn under the apron.

In all centres, cardiologists use thyroid protection, not always used by nurses and technicians.

Furthermore, algorithms employed for estimating E are quite different.

The results presented here reflect the fact that deriving effective dose in such non-uniform exposure remains a serious problem since it is not clear which method and which of the many correction factors or algorithms proposed is able to provide the best estimate for E, as recently underlined by Schultz et al. [18].

4.2 Occupational Doses in Interventional Cardiac Laboratories

Staff doses recorded show a large variability: from 0.5 to 6 mSv/year of effective dose estimated according to the Niklason algorithm.

Table 4 Interventional staff dosimetry survey: dosimetry methods applied in various partner hospitals for cardiologists (Card) nurses and radiographers (Rad)

	Number and Position of dosemeters									
	Over protective apron			Under protective apron			Thyroid protection			
Partner	Card	Nurse	Rad	Card	Nurse	Rad	Card	Nurse	Rad	Algorithm to estimate effective dose
Ireland	2	2	–	1	1	1	Y	Y	Y	–
Italy	1	1	1	1	1	–	Y	Y	Y	1 dose: $E = H_p(10)$ 2 dose: $E = 0.06 \times [Hp(0.07) - H_p(10)] + H_p(10)$
Spain	1	1	1	1	1	1	Y	Y	Y	–
Luxembourg	1	–	–	1	1	1	Y	Y	Y	–
Finland	1	1	1	–	–	–				$H_p(10)$
Greece	1	1	1	–	–	–	Y	Y	Y	$E = 0.03 \times H_p(10)$ (NCRP 22)
Cyprus	1		–	1		1	Y	Y	Y	$E = 0.5 \times H_{under} + 0.025 \times H_{over}$
Slovak Rep.	1	1	1	1	–	–	Y	N	N	$E = H_p(10) + w_{skin} \times H_p(0.07)$
Estonia	1	–	–	1	1	1	Y	Y	Y	–
Slovenia	–	–	–	1	1	1	Y	Y	Y	$H_p(10)$
Bulgaria	–	–	–	1	1	1	Y	Y	Y	–
Scotland	–	–	–	1	1	1	Y	Y/N	N	Unknown

Table 5 Median and third quartile of effective dose and personal dose equivalent over the protective apron for the first operator, evaluated in a sample of European cardiac centres

Annual dose (mSv)	Median	Third quartile
Effective dose	1.3	1.4
Equivalent dose over the apron (H_{over})	11.1	14

The third quartile values can be adopted as constraints for the annual dose for E and H_{over} useful to identify poor practices where an optimisation action is required (Table 5).

Staff monitoring optimisation is required and the following actions are probably necessary:

- To promote the use of the double dosimetry technique, one over the protective apron at the collar level and the second under the protective apron at the chest or wrist level
- To promote studies to identify most appropriate dosimetry methods and algorithms
- To develop a European guideline addressing staff protection system and dosimetry methods in interventional radiology procedures
- To promote the assessment of dose constraints for staff operating in the different interventional area
- To promote dosimetry audits aiming to identify poor radiation protection practices

References

1. Onnasch DG, Schroder FK, Fischer G, Kramer HH. Diagnostic reference levels and effective dose in paediatric cardiac catheterisation. Br J Radiol. 2006 Aug 17.
2. Struelens L, Vanhavere F, Bosmans H, Van Loon R, Geukens M. Data analysis from a multicentre, comparative study of agiographic examinations leading to practical guidelines for the optimisation of patient doses. Radiat Prot Dosimetry 2005;117(1–3):87–92.
3. Dragusin O, Desmet W, Heidbuchel H, Padovani R, Bosmans H. Radiation dose levels during interventional cardiology procedures in a tertiary care hospital. Radiat Prot Dosimetry 2005;117(1–3):231–235.
4. Verdun FR, Aroua A, Trueb PR, Vock P, Valley JF. Diagnostic and interventional radiology: a strategy to introduce reference dose level taking into account the national practice. Radiat Prot Dosimetry 2005;114(1–3):188–191.
5. Brambilla M, Marano G, Dominietto M, Cotroneo AR, Carriero A. Patient radiation doses and references levels in interventional radiology. Radiol Med (Torino). 2004 Apr;107(4):408–418. English, Italian.
6. Aroua A, Rickli H, Stauffer JC, Schnyder P, Trueb PR, Valley JF, Vock P,Verdun FR. How to set up and apply reference levels in fluoroscopy at a national level. Eur Radiol. 2006 Oct 27.
7. Neofotistou V, Vano E, Padovani R, Kotre J, Dowling A, Toivonen M, Kottou S, Tsapaki V, Willis S, Bernardi G, Faulkner K. Preliminary reference levels in interventional cardiology. Eur Radiol 2003 Oct;13(10):2259–2263.

8. Trianni A, Padovani R, Foti C, Cragnolini E, Chizzola G, Toh H, Bernardi G, Proclemer A. Dose to cardiologists in haemodynamic and electrophysiology cardiac interventional procedures. Radiat Prot Dosimetry 2005; 117(1–3):111–115.
9. Delichas M, Psarrakos K, Molyvda-Athanassopoulou E, Giannoglou G, Sioumdas A, Hatziioannou K, Papanastassiou E. Radiation exposure to cardiologists performing interventional cardiology procedures. Eu J Radiol 2003; 48:268–273.
10. Vano E, Guibelalde E, Fernandez JM, Gonzalez L, Ten JI. Radiation exposure to medical staff in cardiac radiology. Br J Radiol 1998; 71:954–960 11.
11. International Commission on Radiation Protection. General Principles of Monitoring for Radiation Protection of Workers. ICRP Publication 75.1997.
12. National Council on Radiation Protection and Measurements. Use of Personal Monitors to Estimate Effective Dose Equivalent and Effective Dose to Workers for External Exposure to Low-LET Radiation. NCRP Report 122.1995.
13. Padovani R, Foti C, Malisan MR. Staff dosimetry protocols in interventional radiology. Radiat Prot Dosimetry 2001; 94(1–2):193-196.
14. Niklason LT, Marx MV, Chan HP. The estimation of occupational effective dose in diagnostic radiology with two dosimeters. Health Phys 1994; 67(6): 611–615.
15. Karppinen J, Parvianen T, Servomaa A, Komppa T. Radiation risk and exposure of radiologists and patients during coronary angiography and percutaneous transluminal coronary angioplasty (PTCA). Radiat Prot Dosimetry 1995; 57:481–485.
16. Axelsson B, Cederlund T, Svane B. Evaluation of radiation exposure to personnel in cardiac angiography. Radiat Prot Dosimetry 1995; 57(1–4): 433–436.
17. Padovani R, Foti C, Malisan MR. Staff dosimetry protocols in interventional radiology. Radiat Prot Dosimetry 2001; 94(1–2):193–196.
18. Williams JR. The interdependence of staff and patient doses in interventional radiology. Br J Radiol 1997; 70:498–503.
19. Faulkner K, Marshall NW. The relationship of effective dose to personnel and monitor reading for simulated fluoroscopic irradiation conditions. Health Phys 1993; 64(5):502–508.
20. Schultz FW and Zoetelief J. Estimating effective dose for a cardiac catheterisation procedure with single or double personal dosemeter. Radiat Prot Dosimetry 2006; 118(2):196–204.

Optimization of Radio Protection in Gynecology Brachytherapy

Nina Samet

Abstract All developments in technology, dosimetry, oncology and mainly quality assurance – aim at reducing the risk of complications to the patient. As the staff is concerned, radiation oncologists, physicists, nursing staff and technicians, the three general principles of radiation protection should be followed: justification, ALARA (As Low As Reasonably Achievable) and individual dose limitations.

Keywords Brachytherapy • Quality assurance and Quality control • Protection ICRU • ICRP

1 Introduction

The arm of Quality Assurance in brachytherapy is to maximize the probability that each individual treatment is administered consistently, accurately and safe. A very important function in HDR and LDR brachytherapy is the correct geometric localization of the applicator, the placement crucially depends on the skill of the radiation oncologist. Following the application procedure it is first the physicist's responsibility to ensure that the treatment is delivered accurately and safe in accordance with the radiation oncologist's prescription. We need to ensure that sources of correct strength and type are accurately positioned in the applicators, as determined form the reconstruction radiographs and treatment planning procedure and treatment delivery process.

N. Samet (✉)
Department of Radiation Oncology, Moldavian Ocological Institute, Chisinau, Republic of Moldova
e-mail: nina_smd@yahoo.com

Y. Lemoigne and A. Caner (eds.), *Radiation Protection in Medical Physics*,
NATO Science for Peace and Security Series B: Physics and Biophysics 1,
DOI 10.1007/978-94-007-0247-9_21,

2 Brachytherapy

Brachytherapy highly differs from external irradiation, mainly because of two technical characteristics.

2.1 Contact

Radioactive sources are directly in contact with the tissue, with a rapid absorbed dose fall off as a function of the distance. The dose distribution is therefore inhomogeneous. The prescribed dose, usually named as the reference dose, corresponds to the minimal dose inside the target volume. A large part of the target volume receives more than two times the dose. The dose inhomogeneity increases with the distance between the sources.

Because of this heterogeneity, normal tissue tolerance is limited and the treated volume must be kept relatively small.

The dose distribution and the geometry of the implant is never identical to the one in the provisional dosimetry.

The dose distribution and the treatment time are therefore calculated from the images (X-ray or CT-scan or MRI) performed after the implant.

The dose distribution can be modified with an optimization, at least if the geometry of the implant is of a good quality.

This optimization is included in the treatment planning system, when a stepping source with either a pulse dose-rate machine or a high-dose-rate machine is used. An optimization is also possible with iridium wires caesium sources, with manual modification, such as different treatment times for each source or length modification of the sources. It should be noticed however, that an optimization would never transform a bad quality implant into a good implant.

2.2 Organ Accessibility

Source implant inside or in contact with the tumour requires organ accessibility. Usual clinical indications for brachytherapy are relatively superficial tumours or tumours located in cavities, such as oral cavity, oropharynx, nasopharynx, bronchus, oesophagus, vagina, uterus, anal canal, rectum and bladder.

The development of CT-scan and MRI have not dramatically modified brachytherapy indications except some examples requiring this type of sophistication, such as a brain glioma.

The use of 3-D images and stereotactic methods have led to the development of either permanent or temporary implants.

3 Conclusion

The optimization process shall include the selection of equipment, the consistent production of adequate diagnostic information or therapeutic outcome as well as the practical aspects, quality assurance including quality control and the assessment and evaluation of patient doses or administered activities, taking into account economic and social factors.

References

1. A Practical Manual of Brachytherapy, by Bernard Pierquin and Ginette Marinello (translated by Frank Wilson, Beth Erickson) Madison, Wisconsin, 1997.
2. The American Brachytherapy Society Recommendations for High-Dose-Rate brachytherapy for Carcinoma of the Cervix. Subir Nag, M.D., Beth Erickson, M.D. and others, J. Radiat. Oncol. Biol. Phys., Vol. 48, No. 1. pp. 201–211, 2000.
3. ICRP Publication 60, International Commission on Radiation Protection. "Radiation protection-1990 Recommendations of the International Commission on Radiological Protection." Pergamon Press, 1991.
4. ICRU, International Commission on Radiation Units and Measurements. "Dose and volume specification for reporting intracavitary therapy in gynaecology". Report 38 of ICRU, ICRU Publications, Bethesda, MD, 1985.
5. The GEC ESTRO Handbook of Brachytherapy. Edited by Alain Gerbaulet, Richard Potter, Jean-Jacques Mazeron, Harm Meertens, Erik Van Limbergen. 2002.

Radiation Protection Optimisation in Interventional Procedures

Nina Samet and Maria Anna Staniszewska

Abstract The regular use of x-ray techniques for the visual control of different types of medical procedures has become a significant source of radiation both to patients and to staff. The exposure to ionising radiation involved from procedure to procedure differs and depends on the type of x-ray equipment used, the skills of the equipment operator, as well as the physical structure and size of the patient. This paper presents a brief description of non-vascular interventional procedures, in terms of their type, x-ray equipment and absorbed dose to the patient. These procedures are performed mainly by physicians, more often not familiar with Radiation Protection rules and therefore, despite the rather low doses received by the staff, the dose to the patient may be locally very high. This is a main reason for the continuous radiation protection education and training of the staff involved with such procedures.

Keywords X-rays • Fluoroscopy • Interventional radiology • Dose

1 Introduction

The term "Interventional procedure" covers all the medical practices performed with the use of x-rays to visualization and control the procedure on the patient. Ordinary diagnostic x-ray examinations are not included in the term.

N. Samet (✉)
Department of Radiation Oncology, Moldavian Ocological Institute,
Chisinau,
Republic of Moldova
e-mail: nina_smd@yahoo.com

M.A. Staniszewska
Medical University, Łódź, Poland
e-mail: amstanisz@o2.pl

Y. Lemoigne and A. Caner (eds.), *Radiation Protection in Medical Physics*,
NATO Science for Peace and Security Series B: Physics and Biophysics 1,
DOI 10.1007/978-94-007-0247-9_22,

Interventional procedures can be divided into two main groups:

- Vascular
- Non-vascular (or extra vascular)

Vascular procedures can be diagnostic or interventional (therapeutic) and Non-vascular are always therapeutic.

Neuroradiological procedures are a special group that consist of both vascular and extra vascular procedures. Most of the vascular procedures are performed for cardiological purposes.

Here we will concentrate on the non-vascular procedures. These cover mainly surgical practices performed under fluoroscopic control, and also drainage, puncture and percutaneous needle biopsy and lithotripsy.

2 Medical Practices Using Fluoroscopic Control (Neurology, Urology, Gastroenterology (GI), Orthopaedics, Genecology and Anaesthesiology)

2.1 Neurology – Non-Vascular Procedures

These procedures concern mainly the spine, for example:

- Localizing and correcting fractures and or injuries
- Reposition of fractures and control
- Embolisation of fistulas
- Vertebroplasty (the vertebral body is repaired by filling it with artificial materials and bone glue. Some times it may be necessary to follow up with angiography using contrast media)
- Discography (entrance into the intravertebral space and endoscopic control)

Most of the exposures during the above procedures are with the patient in the Lateral projection and the x-ray tube over the table.

2.2 GI Procedures

The GI procedures are manifold and their performance depends strongly on the professional experience of the medical teams. These include:

- Percutaneous transhepatic cholangio drainage (PTCD) and stent implantation in the bile duct (performed at the angiographic laboratory)
- Chemonucleolysis and nucleotomy (performed at the surgical suite)
- Endoscopic removal of the gall bladder with laparoscopy under fluoroscopic control with a C-arm unit (performed at the surgical suite)

- The investigation of the bile duct for remaining stone segments (performed at the surgical suite)
- Treatment of GI bleeding by embolisation (performed at the treatment room)
- The placement of oesophagial stents (performed at the surgical suite)
- Endoscopic retrograde cholangio pancreatecography (ERCP) also belongs to interventional procedures because fluoroscopy is required to guide the endoscope
- Biliary Extracorporeal Shock-Wave Lithotripsy (ESWL), where a catheter is positioned in the bile duct. This procedure requires a biplane fluoroscopy system
- Palliative treatment of patients with esophageal cancer. In this procedure the endoscope is under fluoroscopic control for the introduction of the dilatator

2.3 Urology

In Urological practices diagnostic as well as therapeutic procedures are used.

The diagnostic procedures are:

- Uretherography
- Retrograde pyelography
- Descending pyelography
- Nephrography

The therapeutic procedures are:

- Percutaneous nephrostomy (PCN)
- Percutaneous nephrolithotomy (PCNL), where the remaining stone fragments are rmoved through the pyelocalyceal system (the probe is inserted into the ureter under fluoroscopic control. The time taken is longer when the stone is larger than 20 mm)
- Interventional urethrorenoscopy (URSL) with the dilatation of the ureter
- Extracorporeal Shock-Wave Lithotripsy (ESWL), where the lithotripsy of the stone in the ureter is carried out by shock waves (generated by EM, piezoelectric or electrohydraulic pulses). Remaining stone fragments are removed by uretherorenoscopy (endoscope guided under fluoroscopic control)

2.4 Gynaecology

In practice only two procedures are performed in gynecology:

- Hysterosalpingography (HSG) which is a non-vascular procedure routinely performed to find the cause of fertility disorders.

- Uterine artery embolisation which is performed with the assistance of interventional radiologists to treat women with emergency uterine bleeding. This is a vascular procedure, known since the 1970s, during which the physician guides a small angiographic catheter into the uterine arteries and injects a stream of tiny particles that decrease blood flow to the uterus. It is now considered a safe and highly effective non surgical treatment of women with symptomatic uterine fibroid tumours. Uterine fibroid embolisation has several advantages over conventional hormonal suppression and surgical procedures.

2.5 *Orthopaedics*

In orthopaedics fluoroscopy is used strictly to control correctness of the surgery reconstructions of bones and skeleton elements, i.e. fracture repositions with metal screws or rods (mostly concerning extremities).

2.6 *Anaesthesiology*

Fluoroscopic control is used during implantation of subcutaneous central entrance for long term application of drugs (Implant types: PORT, Broviac, etc).

3 Requirements for X-ray Equipment Used in the Procedures Listed Above

For non vascular interventional procedures the basic equipment requirements are an advanced stable or mobile C-arm fluoroscopic system equipped with a DAP meter (obligatory), a CP generator and additional beam filtration (Cu is preferred). The system must include pulsed fluoroscopy mode capable of last image hold. The X-ray tube must be designed for long fluoroscopy times at high loads (good heat dissipation) capable to examine thick patients at the Lateral position.

C-arm units have to allow for a good access to the patient from all sides (particularly to the patient's head by the anaesthesiologist).

Some GI procedures require equipment of a higher class, for example:

- PTCD – requires an angiographic system
- ESWL – requires a biplane system because optimal orthogonal projections are required

Mobile C-arm units not designed for interventions are used frequently in orthopaedics, for example an over couch system with a highly absorbing table, without pulsed fluoroscopy, without DAP meter and low quality of image intensifier. Unfortunately such equipment is used by orthopaedic surgeons not familiar with radiation protection and x-ray imaging techniques.

4 Patient Radiation Doses

During interventional procedures the dose to the staff (especially the operators) is strongly dependent on the dose to the patients. Therefore any reduction to patient exposure is a direct reduction to staff exposure.

Interventional radiology (fluoroscopically-guided) techniques are being used by an increasing number of clinicians, who are unfortunately not adequately trained in radiation safety or radiobiology (cardiologists, neurosurgeons, neurologists, orthopaedists, urologists, vascular surgeons). This inadequate training is evident from the DAP values and fluoroscopy time that show an increasing patient exposure to radiation with time.

For a more accurate estimation of radiation risk, measurement of the entrance dose (entrance air kerma) is required with the evaluation of the absorbed dose to individual organs. It should be noted, that the effective dose is a good measure of radiation risk when the exposure is nearly uniform, or it covers a big area of the body. Therefore in interventional procedures where the primary beam exposes only the head or only the extremities, the effective dose can be low although the local surface dose is very high (even erythema is locally possible).

Examples of doses to patients during interventional procedures are given below from studies carried out in different countries. These have to be treated as estimation of the order of magnitude, because exposure parameters (and thus intensity of primary x-ray beam) are automatically selected according to the body size of an individual patient and can highly differ for the same procedure with different patients.

4.1 Italy, 2004 [1]

DAP in nephrostomy and percutaneous transhepatic cholangiography: (60–160) Gy cm^2. Cystourethrographies and ERCP: 25 Gy cm^2. For vascular procedures of the lower limbs: approximately 100 Gy cm^2. For vascular procedures of the abdomen: approximately 450 Gy cm^2.

Mean effective dose estimated for patients:

- Extravascular procedures between 5 and 28 mSv
- Neuroradiological between 13 and 33 mSv
- Vascular procedures involving the abdomen between 36 and 87 mSv

Correlation between total fluoroscopy timc and DAP values was poor.

4.2 The Netherlands, 2002 [2]

Neurointerventional vascular procedures (can involve very high doses to the patient) for Aneurysm, embolisation of arteriovenous malformations the entrance dose to the skull of the patient: the highest was 2.3 Gy, the average was 0.9 ± 0.5 Gy. The effective dose to the patient was estimated as 14.0 ± 8.1 mSv. The average DAP measurement was 228 ± 131 Gy cm^2. The average fluoroscopy time was 34.8 ± 12.6 min. The highest effective dose to the operator during these procedures was approximately 7 μSv.

4.3 Sweden, 2009 [3]

Percutaneous vertebroplasty (PVP): The mean effective dose to patients was 12 mSv. The effective dose to the operator was <1 mSv (at a particular position of the x-ray tube).

4.4 USA, 2009[4]

Hysterosalpingography (HSG) with the use of a mobile C-arm fluoroscopy unit:
Mean estimated surface dose for:

- A normal patient size was 2.6 mGy
- An abnormal patient size was 6.9 mGy
- Selective catheterization (for fallopian tube occlusion) was 46.7 mGy

The mean fluoroscopy time was from 4 up to 56 s.

References

1. Brombilla M, Marano G, Dominietto M, et al.: Patient radiation doses and reference levels in interventional radiology. Radiol Med., 2004, 107(4): 408–418.
2. Kemerink GJ, rantzen MJ, Oei K. Sluzewski M, et.al: Patient and occupational doses in neuro-intervenional procedures. Neuroradiology, 2002, 44(6), 522–528
3. Von Wrangel A, Cederblad A, Rodriguez-Catarino M. Fluoroscopically guided percutaneous vertebroplasty- assessment ofradiation doses and implementation of procedural routines to reduce operator exposure. Acta Radiol., 2009, 50(5): 490–496
4. Phillips J, Cochavi S, Silberzweig JE, et al: Hysterosalpingography with use of mobile C-arm fluoroscopy. Ferti Steril, 2009, Feb. 5 (Epub ahead of print)

Radiation Protection Optimisation in Diagnostic Nuclear Medicine

Stelios Christofides

Abstract The ultimate goal of any type of medical imaging procedure is to obtain the best image quality while delivering the smallest radiation dose possible to the patient. The best image quality though, does not necessarily give the correct diagnosis for a given medical condition at the lowest possible dose to the patient. Additionally the vast number of alternative diagnostic modalities available today and their rapid evolution make the choice of the most suitable modality for a particular medical condition very difficult, if dose to the patient is to be considered as a major constraint. It is therefore very important to know the dose received by the patient from the different modalities to arrive at the same diagnostic result. This is especially important in Nuclear Medicine where the different modalities produce images of the metabolic function of the human body and they are more likely to arrive at the same diagnostic outcome. The aim of this presentation is to give an overview of the methods used to optimise the diagnostic value of the images produced by Nuclear Medicine diagnostic modalities.

Keywords Nuclear medicine • Quality control • Optimisation • Radiation protection

1 Introduction

The diagnostic value of each Nuclear Medicine diagnostic procedure depends on the quality of the radiopharmaceutical administered to the patient and the efficient functioning of the instrumentation used to produce an image, a series of images or other dynamic function curves that will assist the Nuclear Medicine Physician to make a diagnostic evaluation of the patient.

S. Christofides (✉)
Medical Physics Department, Nicosia General Hospital, Strovolos, Nicosia, Cyprus
e-mail: cstelios@cytanet.com.cy

Y. Lemoigne and A. Caner (eds.), *Radiation Protection in Medical Physics*,
NATO Science for Peace and Security Series B: Physics and Biophysics 1,
DOI 10.1007/978-94-007-0247-9_23,

The factors affecting the diagnostic value of a Nuclear Medicine diagnostic procedure are:

1. The Equipment and other instrument used
2. The Quality of the Pharmaceutical used
3. The procedure used
4. The patient Physiology

In this presentation 1 and 2 and some element of three will be elaborated further since these factors are directly related to the diagnostic value of the Nuclear Medicine procedures used to obtain the required data from the patient.

2 The Equipment and Other Instruments Used

The choice of a procedure will depend on the availability of the modalities as well as the radiopharmaceuticals in a particular Nuclear Medicine Centre. The justification for choosing, from the available procedures, one procedure than the other, is the responsibility of the Nuclear Medicine physician, who must have the necessary education and experience to make the correct judgment for the particular patient under investigation.

The correct functioning of the equipment and other instrumentation, such as dose calibrators, must be assured and a Quality Assurance programme is essential. The Quality Control tests that are performed on regular time intervals depend on the type of equipment or instrument to be tested. The Technologists perform some of the required quality control tests and the Medical Physicist performs others. Table 1 gives an example of such tests for a SPECT system.

Table 1 Quality control tests for a SPECT system

QC Test	Acceptance	Daily	Weekly	Yearly
Uniformity	P	T	T	P
Uniformity tomography	P			P
Spectrum Display	P	T	T	P
Energy resolution	P			P
Sensitivity	P		T	P
Pixel Size	P		T	P
Centre of Rotation	P		T	P
Linearity	P			P
Resolution	P			P
Count losses	P			P
Multiple window positions	P			P
Total performance phantom	P			P

P = Physicist, T = Technologist

3 Current Strategies

The administration of a radiopharmaceutical or radionuclide to a patient renders the patient a moving radioactive source. Furthermore the quality of the administered radiopharmaceutical needs to be assured in order to safeguard the patient form adverse effects and unnecessary exposure. Therefore the movement of the patient within the Nuclear Medicine department as well as the quality of the radiopharmaceuticals administrated to the patient are major factors affecting the dose to the patient. These are minimised by considering:

1. The design of a Nuclear Medicine Department
2. The Quality of the Radiopharmaceutical administrated to the patient
3. The Amount of Radiopharmaceutical Administrated

3.1 The Design of a Nuclear Medicine Department

This requires special arrangements in order to minimise the exposure of other patients, the staff as well as the patient from other administrated patients. This necessitates special requirements to be taken into account in the design of a Nuclear Medicine Department that will restrict and minimise such exposures from taking place.

3.2 The Quality of the Pharmaceutical Used

The quality of the radiopharmaceutical administrated to the patient has to be such that it will not cause any adverse effects to the patient, expose the patient to unnecessary radiation and at the same time be specific for the organ of interest. It is the responsibility of the manufacturer or the supplier of the radiopharmaceutical to make sure that the following parameters are within the acceptable limits:

- Radionuclide purity (other radionuclides present?)
- Radiochemical purity (labelling efficiency)
- Chemical purity (toxic substances present?)
- Sterility
- Absence of pyrogens

Although the above parameters are the responsibility of the manufacturer or the supplier, nevertheless these should also be checked at the Nuclear Medicine Department before the radiopharmaceutical is administered to the patient. This forms part of the radiopharmaceutical quality assurance programme that consists of the following steps:

- Quality control of radiopharmaceuticals
- Written and trained procedures in preparation and safe handling of radiopharmaceuticals
- Use of unique code that guarantee the ability to trace the origin of all components in the preparation
- Records of radionuclides, kits, etc
- Labelling of vials and syringes
- Measurement of activity

Before injecting the patient with the radiopharmaceutical the following must be verified:

- Patient name
- Patient identification number
- Is the patient pregnant?
- Is the patient breast-feeding?
- Check the request form
- Check the label of the syringe. Is it the correct radiopharmaceutical for the requested examination?
- Is it the correct activity for the physiology of the patient?

If the above are all correct then the patient is injected with the radiopharmaceutical.

3.3 *Amount of Radiopharmaceutical Administered*

As it was mentioned in a previous presentation (Patient Dose Assessment), the administrated amount of radiopharmaceutical depends on the patient's:

1. Gender
2. Age
3. Height
4. Weight

Also the maximum amount administered should not exceed the maximum allowed organ dose. As the injected radiopharmaceutical circulates in the blood system before it is absorbed and preferentially concentrate in the organ of interest, other organs of the body absorb some of the radiopharmaceutical and therefore receive a dose proportional to the amount of radiopharmaceutical absorbed, taking into account the composition of the organ and the type of radiopharmaceutical.

The required activity to be administered is given by the manufacturer of the radiopharmaceutical in the leaflet accompanying the radiopharmaceutical.

4 Recent Advances and Challenges

4.1 Molecular Imaging

In the recent years the term "Molecular Imaging (MI)" is extensively being used. It is appropriate at this stage to explain this term and also the challenges that it brings with it.

4.1.1 Definitions of Molecular Imaging

MI is the visualisation/characterisation of biological processes in living organisms. MI techniques directly or indirectly monitor and record the spatio-temporal distribution of molecular or cellular processes for biochemical, biological, diagnostic or therapeutic applications. Furthermore MI is the non-invasive, quantitative and repetitive imaging of targeted macromolecules and biological processes in living organisms.

Therefore MI is the visual representation, characterisation and quantification of biological processes at the cellular or sub-cellular levels within intact living organisms. In other words MI is the studying of diseases non-invasively at the molecular level. MI is an interdisciplinary approach involving biologists, physicists, physicians, mathematicians, conventional chemists, radio-chemists and other specialists who have joint forces for the better understanding and visualisation of the normal physiological processes and the molecular processes preceding the morphological manifestations of diseases in-vivo.

Taking the above definitions into consideration one may conclude that Nuclear Medicine is a sub-Specialty of Molecular Imaging.

4.1.2 Expectations

MI is expected to be pivotal to early diagnosis, patient stratification and early response assessment. MI is aimed at the characterisation, visualisation and quantification of specific cellular and molecular processes underlying diseases.

The convergence of technology in recent years within different fields of research has broadened the horizon for MI applications both in basic research and the clinical setting. These include and are not limited to:

- The mapping of the entire human genome
- Tools to genetically manipulate cells or organisms
- New treatment strategies, including molecular and cell based therapies
- Nonotechnology
- State-of-the-art imaging modalities with increased spatial and temporal resolutions

In the setting of personalised medicine, the exact molecular background of a disease in a single patient could be assessed and an individualised treatment regiment could be designed and therefore maximise the therapeutic effect and minimise adverse effects.

Currently the most prominent or pressing clinical applications of MI are in the fields of Oncology and Cardiovascular diseases. In Oncology, MI has already resulted in some breakthroughs in cancer staging and treatment response assessment. It is anticipated to facilitate early disease detection. In Cardiovascular disease, MI is expected to serve a central role in addressing basic questions regarding cell therapy for cardiac repair and vulnerable plaque detection.

4.2 Hybrid Systems

Medical imaging technology has evolved rapidly in the last couple of decades, where today, detailed three-dimensional images of the body can be obtained in a few seconds. The radiation doses delivered to the patient by some of these sophisticated procedures are considerable but so are the benefits. In contrast, the radiation doses associated with the majority of the routine imaging examinations involving conventional radiography and fluoroscopy are gradually reducing, due to the developing technology and the increasing sensitivity of the evolving imaging devices.

Computed Tomography (CT) practice continues to evolve with the introduction of multi-slice (MSCT) that utilises multi-detector rows (MDCT) to allow fast helical scanning and rapid imaging of large volumes of the patient. Such technology is promoting the further development of new and complex diagnostic and interventional CT procedures, with clear potential for increasing doses to individuals and the population. Conversely, there is increasing attention to optimisation of patient protection through improvements in CT technology and practice with some possibilities for dose reduction.

Over the past 15 years there has been substantial changes in Nuclear Medicine techniques, for example the routine application of Single Photon Emission Computed Tomography (SPECT) imaging and the introduction of Positron Emission Tomography (PET) in clinical Practice.

Today, dedicated Positron Emission Tomography (PET) systems are the most universally installed systems. Mobile scanners and modified gamma cameras (CGPET) are used occasionally as a lower cost alternatives and interest in PET-CT hybrid systems is rising despite the limited assessment of impact on service planning. PET was used and assessed most commonly for managing patients with cancer.

4.3 New Pharmaceuticals

Different radiopharmaceuticals are used depending on the Molecular Imaging modality used (PET or SPECT). Also for a specific examination there may be more that one radiopharmaceutical that can be used to acquire the final image. The dose required to produce the final image at the required quality, depends on the radiopharmaceutical used as well as on the time constrains imposed on the image acquisition.

Currently there is a large effort in the development of new PET tracers, as well as SPECT for Molecular Imaging. The development of such tracers, render the use of SPECT/CT and dynamic SPECT systems capable of Molecular Imaging in areas such as cardiology. For example they may be used on the imaging of rupture prone, vulnerable plaque with agents such as annexin V9 or matrix metalloproteinases or integrins, labelled with Technitium-99 m, Iodine-123, or Indium-111. The need for precise anatomic localisation of the tracers that are concentrated in the vulnerable plaques is clear for such applications.

Another example that in some countries is already used in the clinical practice for the differential diagnosis of Parkinson Syndrome and Essential Tremor is the radiopharmaceutical DaTSCAN with a SPECT system.

5 Summary

The Medical Exposures in Nuclear Medicine examinations and effectively the radiation protection optimisation depends on the following:

- Choice of examination
- Quality control of the equipment and the radiopharmaceuticals
- Optimisation of the administered activity
- Safe routines to avoid miss-administration of the radiopharmaceuticals
- Methods of reducing the absorbed dose
- Quality assurance of the procedures and methods used

Acknowledgements The material used for this presentation, have been taken from the following sources:

- The SENTINEL European Project, www.dimond3.org
- IAEA, "Nuclear Medicine Resource Manual", STI/PUB/1198, IAEA, Austria, 2006
- The IAEA Nuclear Medicine Educational and Training material available at (http://rpop.iaea.org/RPoP/RPoP/Cotent/index.htm)
- The EANM web site (www.eanm.org)
- The ESMI web site (www.e-smi.eu)

These are freely accessible and you are encouraged to study and use them.

Appendix: Color Section

Abstract This section is devoted to figures from different papers of this book which needs to be printed in color instead of black and white like in the previous chapter. In the chapters, these figures have a note: "See color picture in Appendix 1". In this appendix, they are recognized by the author name and the figure number they had in the chapter.

1 P. Peschke

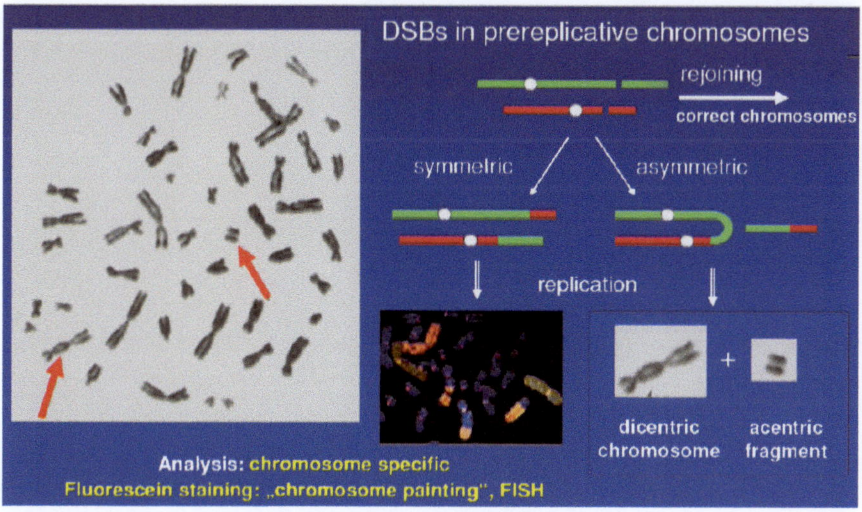

Fig. 4 Chromosomal aberrations after mis-repair of double strand breaks

Y. Lemoigne and A. Caner (eds.), *Radiation Protection in Medical Physics*, NATO Science for Peace and Security Series B: Physics and Biophysics 1, DOI 10.1007/978-94-007-0247-9, © Springer Science+Business Media B.V. 2011

2 Kukolowicz

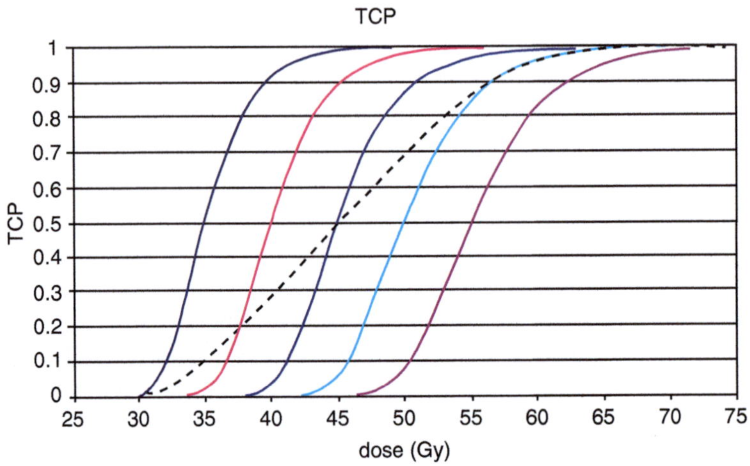

Fig. 3 The TCP curves for each homogenous group of patients separately and for the whole group (*dotted line*)

MIX
Papier aus verantwortungsvollen Quellen
Paper from responsible sources
FSC® C105338

If you have any concerns about our products,
you can contact us on
ProductSafety@springernature.com

In case Publisher is established outside the EU,
the EU authorized representative is:
Springer Nature Customer Service Center GmbH
Europaplatz 3, 69115 Heidelberg, Germany

Printed by Libri Plureos GmbH
in Hamburg, Germany